The Complete Smithsonian Field Guide

American Missiles
1962 to the Present Day

The Complete Smithsonian Field Guide

American Missiles
1962 to the Present Day

By Brian D. Nicklas
National Air and Space Museum
Smithsonian Institution, 2011

Frontline Books

American Missiles: 1962 to the Present Day

This edition published in 2012 by Frontline Books,
an imprint of
Pen & Sword Books Limited,
47 Church Street, Barnsley, S. Yorkshire, S70 2AS
www.frontline-books.com

Email info@frontline-books.com or write to us at the above address.

Copyright © Brian D. Nicklas, 2012

ISBN: 978-1-84832-517-3

The right of Brian D. Nicklas to be identified as the author of this work has been asserted by him in accordance with the Copyright, Designs and Patents Act of 1988.

All rights reserved. No part of this publication may be reproduced, stored in or introduced into a retrieval system, or transmitted, in any form, or by any means (electronic, mechanical, photocopying, recording or otherwise) without the prior written permission of the publisher. Any person who does any unauthorised act in relation to this publication may be liable to criminal prosecution and civil claims for damages.

CIP data records for this title are available from the British Library

Typeset in 10pt Baskerville by
Mac Style, Beverley, E. Yorkshire

Printed and bound in India by Replika Press Pvt. Ltd.

Pen & Sword Books Ltd incorporates the Imprints of Pen & Sword Aviation, Pen & Sword Family History, Pen & Sword Maritime, Pen & Sword Military, Pen & Sword Discovery, Wharncliffe Local History, Wharncliffe True Crime, Wharncliffe Transport, Pen & Sword Select, Pen & Sword Military Classics, Leo Cooper, The Praetorian Press, Remember When, Seaforth Publishing and Frontline Publishing.

For a complete list of Pen & Sword titles please contact
PEN & SWORD BOOKS LIMITED
47 Church Street, Barnsley, South Yorkshire, S70 2AS, England
E-mail: enquiries@pen-and-sword.co.uk
Website: www.pen-and-sword.co.uk

Contents

Introduction	vi
Acknowledgements	xiii
About the Author	xiii
Numerical List of Missiles	xiv
American Missiles	1
Missile Name List	162
Missile Families	166
Bibliography	167
Missiles held by the National Air and Space Museum	170
Index of Designations, Names and Acronyms	171
Photo Credits	175

Introduction

A missile is a projectile, and a projectile is an object propelled through the air (or space) by thrust. Thrust may be applied by an outside force such as a rubber band (as with a slingshot) or even by throwing with one's arm. Or, propulsion might be by an internal motor. A motor may be an air-breathing type, such as jet engine (including a turbojet) or ramjet or pulsejet, or, it could be a rocket – either solid-fuel or liquid-fuel, or could be boosted into the air by a rocket, solid or liquid. Projectiles can be the object of focus in a sport, such as a golf ball or football, or from debris cast out by a shattering window, but the term projectile in most cases is used to describe a weapon. As a weapon, a projectile could be an artillery shell fired from cannon, an unguided, or a guided missile. Here we look at the guided missile.

Simply put – in fact to over-simplify – a missile is a tube pointed at one end and open at the other for rocket exhaust (if the missile is rocket propelled). But perhaps the earliest pyrotechnicians had to explain to the Chinese emperors the differences between short- and long-range missiles, or even between military rockets and celebratory fireworks. Did senior officers in the military arms of the 19th Century and journalists of the period need Congreve and Hale rockets explained to them? (For those who do not know, Sir William Congreve and William Hale were the two most prominent British rocket pioneers in the 19th century.) Perhaps when the "rocket's red glare" illuminated the night sky during the bombardment of Fort McHenry in Baltimore during the War of 1812, the various versions of the British Congreve rockets were too small in number for any party to be concerned. Now, in the 21st Century, missiles are varied in size, shape, speed and purpose, many with and some without an explosive warhead. When launched from the aerial and submarine platforms now in use, missiles are used from environments outside the comprehension of the military leaders of ancient times.

Rather than an individual history of each and every weapons system or aerial vehicle to be bestowed with an "M for Missile" designation, this volume hopes to be a visual guide to these devices. A full history for some could easily be crafted on a single page; other programs would need an entire volume on their own. In 1962, a new system of identifying aerospace military hardware came into place for the American military. This nomenclature combined the aircraft and missiles of the Army, Navy and Air Force under common terms, to ease identification. In the 1990s, as the Archives Division staff, National Air and Space Museum (NASM) of the Smithsonian Institution organized and rehoused the Space Reference Files an effort was made to use the military system in place to identify appropriate files under the same terminology. A few years later, the Museum acquired through a generous donation, the Herbert S. Desind Collection. It was while indexing, processing and re-housing the Desind Collection at the National Air and Space Museum's Archival Support Center at the Paul E. Garber Facility, that this author became frustrated by having loose images of missiles that were not properly identified. It seemed that information on a good number of missiles had not been held previously by NASM. As I looked about for a reference tool to assist me, this book was born. There were a few places on the internet I could search, but I was not near a computer to access those and needed something I could physically have at hand. Watching the television news coverage of various military operations and reading about these same events at the same time, and seeing missiles misidentified also reinforced the idea that such a publication would be useful. As new missiles come into development and use, the confusion that seems prevalent in assigning a designation at all also indicates that a single volume on the topic will be an aid to those working on the subject at hand. Some of the drones that appear in this publication may seem out of place. This may explain why in the late 1990s drones were renamed "unmanned aerial vehicles" and given a "Q" rather than "M" prefix.

A few words about the major source of images for this publication. Herbert S. Desind was a high school science teacher in the Washington DC area with a passion for rocketry and spaceflight. He collected images from all over the world to assist with his teaching in the classroom and with his hobby of making model rockets. Before his death, he acquired over 100 cubic ft of images, not only of manned spacecraft, satellites and airplanes, but also of modern American military missiles, and those make up the bulk of the images used in this publication. Other images were added by contacting the military services or the National Archives for older missiles, and the military and defense contractors for the recent missiles.

The reader will note that the date spread of this work is 1962–2011. These dates as chosen as the start of the current Military Designation System to the present, and does not include earlier missiles that e were not re-designated. The reader should know too, that this is not an historical work and therefore does not go into the developmental histories of the missiles covered in the work but is strictly a ready directory of all known U.S. missiles within this time frame. For those who wish to learn more about the histories of some of these missile systems they need consult some of the sources in the Bibliography in the back of the book.

The missiles are arranged in sequence by their "M" number. Much as American military aircraft have an alpha-numeric designator showing the type and identity, such as the F-16 fighter or the KC-135 tanker, missiles have designations as well. All should have designations rather than names when they enter service, *should* being the key word. And much as the NKC-135A designation will let you know that the variant of the Stratotanker you are looking at is a permanently modified tanker version of the first cargo aircraft from the -135 series, the missile designation ASM-135A lets you know you are looking at a first version anti-satellite missile, the 135th in the missile series.

The US Army, US Navy and US Air Force all had different ways to designate their aircraft and missiles until the fall of 1962 when a joint system came into use. While some missiles were due to be withdrawn from service yet were given new designations; others were just entering service with two or more military branches and were given one designation to enable procurement and identification with greater ease. Some missiles had earlier been given aircraft designations, the 1962 system had them change to M–for-missile designations. The standard which appears to have been applied to the subject of this volume is that guided weapons are missiles, unguided weapons are rockets given an "R" designation, such as the GAR-2 "Genie" rocket fired from F-106 interceptors.

In some cases, this publication will have a blank where no information is known. This may be a case where the missile in question is still classified; it may be a case where the designation was left open for a project that never came to fruition. There are instances where a designation was assigned to a paper project, with little or no hardware built. Some missile contracts were competed for, with several firms building a weapon, with only the winner being assigned a military designation. Other times both missiles for one contract had designations assigned, but only one would see production. What complicates matters is when a weapon is produced and fielded seemingly without being assigned a number at all. This is the case with the current Ground-Based Interceptor (GBI) which has at least been launched in tests, but has yet to receive a designation. Some designations, such as that of the ASM-135, are totally overlooked as the popular name takes hold and is promulgated by the media and others. The ASM-135 is commonly called the ASAT, which is as useful as calling the E-3 Sentry the AWACS, when the Sentry is but one example of an Airborne Warning and Control System aircraft. Currently the ASM-135 is the only ASAT missile, but that may change in years to come. The Tomahawk cruise missile is not the only American cruise missile, yet using the name rather than the designation brings to mind the problem with all small light aircraft being called "Cubs", not just the Piper J-3. The BGM-109 Tomahawk competed with the BGM-110, which although built and flown, did not win a production contract.

What would seem to be a straightforward system devised to end confusion has enough exceptions to cause more confusion, not lessen it. A missile designation would seem to imply a rocket, or at the very least a jet-powered vehicle, yet not only are there missile designations applied to drones with small (in some case single-cylinder) reciprocating engine, there are missiles that are unpowered, being more aptly guided bombs such as the AGM-62. An example of where this was rectified is the case of the AGM-112, which was re-designated as the GBU-15.

There are enough drones on the list to ask the question of why it took until the later 1990s before drones (once RPVs or Remotely Piloted Vehicles, now UAVs or Unmanned Aerial Vehicles) were given their own category, now becoming the "Q" series. There had been reluctance across the board to elevate the importance of UAVs in the military, as supporters (likely to their own detriment) have proclaimed each new era of UAVs as the beginning of the end of manned combat aircraft. This shift can be noted in that the BQM-155 was re-designated the RQ-5.

Successive improvements to a design would be indicated by a change to the suffix letter after the missile number. The liquid-fueled CIM-10A Bomarc became the revised CIM-10B with a solid-fuel engine installed. Yet there are exceptions to this, or what would appear to be aberrations. The BQM-34 Firebee underwent a distinct physical change to become the Firebee II, yet it remained with the BQM-34 designation as it was felt that the guidance and control systems had an over-riding degree of commonality with the not-so-sleek earlier versions. An exception to this are that several of the Radioplane drones are so common between each other that the use of an "A" or "B" suffix would seem to suffice, yet the designations of MQM-33 and MQM-36 were applied. One should think the only difference was between being bought for the Air Force or the Navy, and that one number would be enough.

This volume focuses on designated missiles. Programs flown, yet undesignated are omitted. Previous missiles that did not survive the "rebranding" of designations in 1962 are also omitted. There are several examples of designated missiles for which only the designation and the cursory information that provided are listed. These can be missiles for which the requirement evaporated after the numbers were assigned, with no drawings or fabrication taking place. Or the minimal designation may hold the place for a missile that may have been required by one branch of the military, but had actually duplicated the efforts of another missile in use by another branch. The duplication may have been service vs service, or a case of an older upgraded missile matching the capabilities of the proposed new weapon system. All the information found for this publication came from open sources, so some of the "blank" designations may be for missiles that were indeed built, but where any details remain classified at some level. Given the nature of the subject, it is hoped that this would be a "first edition" on American Missiles, and that omissions, revisions, corrections and updates would follow, to include information from those that may have worked on a particular program and are now available to provide information on what may have been more secretive in the past. There are a good number of missiles for which I was unable to obtain an image. It would appear that unsuccessful programs disappear quickly from corporate records, and often for a missile under initial testing, those records would not be found in Department of Defense holdings.

As a visual guide to these devices, it should be noted that details may be lacking in some images. With 174 designations, and trying to obtain an average of two images per designation, the number of images became daunting. While the AIM-9 Sidewinder has not changed very much in general, the BQM-34 Firebee has undergone enough visual changes where more than two images were required. But please note that if one were aiming to use these images for scale modeling purposes both large and small, you will likely not find the desired amount of details. Again, that would be best left to a single volume on that particular missile.

The omission of scale drawings is a result of the lack of such information for the majority of the subjects of this book. While the National Air and Space Museum's Archives Division holds over 71,000 technical manuals and manual supplements, most of these deal with aircraft, not missiles, not even the missiles that are part of the Museum's collection. Most of the descriptive artwork used in the literature that makes up NASM Archives technical files uses conceptual artwork, not scale plans, so with the lack of a technical manual, presenting a scale three view plan for each subject was impossible.

Finally, while this publication is produced under the auspices of the National Air and Space Museum, Smithsonian Institution, any errors, faults and omissions are due to the author, and should in no way be construed as the official posture of the National Air and Space Museum, the Smithsonian Institution, or the US government, for which the Smithsonian Institution is a trust instrumentality.

The following list outlines the symbols used in guided missile, rocket, probe, booster and satellite MDS (Mission Design Series) designations. After this table are samples of missile MDS designations.

Status Prefix		Launch Environment		Mission		Vehicle Type	
C	– Captive	A	– Air	C	– Transport	B	– Booster
D	– Dummy	B	– Multiple	D	– Decoy	M	– Guided Missile
J	– Special Test (Temporary)	C	– Coffin	E	– Electronic/ Communications	N	– Probe
M	– Maintenance	F	– Individual	G	– Surface Attack	R	– Rocket
N	– Special Test (Permanent)	G	– Surface	I	– Aerial/Space Intercept	S	– Satellite
X	– Experimental	H	– Silo Stored	L	– Launch Detection/ Surveillance		
Y	– Prototype	L	– Silo Launched				
Z	– Planning	M	– Mobile	M	– Scientific/Calibration		
		P	– Soft Pad	N	– Navigation		
		R	– Ship	Q	– Drone		
		S	– Space	S	– Space Support		
		U	– Underwater	T	– Training		
				U	– Underwater Attack		
				W	– Weather		

Sample MDS for Missile Series:

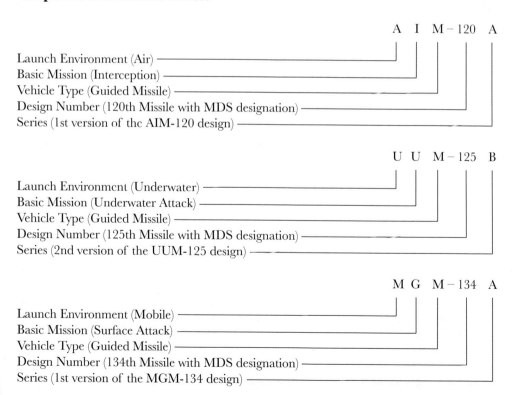

A I M – 120 A

Launch Environment (Air)
Basic Mission (Interception)
Vehicle Type (Guided Missile)
Design Number (120th Missile with MDS designation)
Series (1st version of the AIM-120 design)

U U M – 125 B

Launch Environment (Underwater)
Basic Mission (Underwater Attack)
Vehicle Type (Guided Missile)
Design Number (125th Missile with MDS designation)
Series (2nd version of the UUM-125 design)

M G M – 134 A

Launch Environment (Mobile)
Basic Mission (Surface Attack)
Vehicle Type (Guided Missile)
Design Number (134th Missile with MDS designation)
Series (1st version of the MGM-134 design)

x American Missiles

To keep things simple in this book, the three letter MDS identifier is as shown by:

1st – Rocket/Missile Launch Environment
2nd – Rocket/Missile Mission
3rd – Vehicle type (here it is the "M for Missile")

(Note that the launch environment letter can be preceded at times by a "Status" letter such as C for captive, X for experimental, Y for prototype, etc; example "XAIM-9" or "ZBGM-110.")

Status
C = Captive. The missile is not launched, used for aerodynamic studies or systems tests.
D = Dummy. The missile is a dummy used for display, retaining the outward look and shape of the operational missile.
J = Special Test (Temporary). The missile has been modified from original configuration, for testing, but can be re-configured to the original use.
M = Maintenance. The missile is use for maintenance practice.
N = Special Test (Permanent). The missile has been modified from original configuration, for testing, and can not be re-configured to the original use.
X = Experimental. The missile is a test version.
Y = Prototype. The missile is a pre-production version.
Z = Planning. The missile will likely be a mockup. Example is the ZBGM-110 missile

Rocket/Missile Launch Environment
All rockets and missiles contain a symbol to indicate the launch method, be it from the air, ground, sea, etc. The following are the currently authorized symbols for launch environments. These are not used for other aerospace vehicles.

A = Air-launched. The missile is launched from an airborne vehicle. Example is the AIM-9 Sidewinder dogfighting missile.
B = Multiple. The missile can be launched from various environments. Example is the BGM-109 Tomahawk missile
C = Coffin. The missile is stored and launched from a horizontal container (less than 45 degree angle) at ground level. Coffin launchers may be either on land or at sea. Example is the CIM-10 Bomarc missile.
F = Individual. The missile is launched by an individual soldier in the field, otherwise referred to as man-portable. Example is the FIM-92 Stinger missile.
G = Ground. The missile is launched directly from the ground surface, including runways. Example is the GQM-98 Compass Cope R.
H = Silo-Stored. The missile is launched from a vertical position in a vehicle, but not buried. Example is the HGM-25 Titan missile on a truck trailer.
L = Silo-Launched. The missile is launched from its storage silo, buried in the ground. Example is the LGM-30 Minuteman missile
M = Mobile. The missile is launched from a mobile ground vehicle. Example is the MIM-9 Hawk missile
P = Pad. The missile is launched from a protected or semi-protected facility at ground level. Example is the PGM-11 Redstone missile.
R = Ship. The missile is launched from a ship or barge. Example is the RIM-116 RAM missile
S = Space. The missile is launched from a spacecraft.
U = Underwater. The missile is launched from a submarine or underwater device. Example is the UUM-44 Subroc missile

Rocket/Missile Mission Symbol

Rockets and missiles are assigned a single mission symbol, which usually denotes the intended target type of the missile. For most types of missile, the combination of launch environment and mission symbols form a launch environment-to-mission combination (surface-to-air, ship-to-submarine) that gives one a good idea of the potential uses for the missile. Some designations remain to be used, as far as determined; no cargo missiles have yet been developed.

C = Transport. Applies to vehicles designed to carry cargo and deliver it to a location. This can also be used to designate a carrier for electronics or weapons systems.
D = Decoy. Applies to vehicles that function as decoys for defeating enemy anti-aircraft and anti-missile defenses. Example is the ADM-20 Quail missile.
E = Electronics. Applies to vehicles that carry out electronic missions such as communications or countermeasures. Example is the CEM-138 Pave Cricket missile.
G = Ground. Applies to vehicles designed to attack surface targets, including vehicles. Example is the AGM-65 Maverick missile.
I = Intercept. Applies to vehicles designed to attack aerial targets, including both aircraft and missiles. Example is the AIM-7 Sparrow missile.
L = Launch detection. Applies to vehicles designed to detect launch of missiles and track and identify enemy aircraft and missiles. This also applies to detection and monitoring of space launches and re-entry.
M = Scientific. Applies to vehicles designed to collect scientific data.
N = Navigation. Applies to navigational assistance designs.
Q = Drone. Applies to a vehicle designed to be remotely controlled. Example is the AQM-81 Firebolt missile.
S = Space support. Applies to vehicles designed to support space programs and activities.
T = Training. Applies to training designs. Example is the ATM-84 Harpoon training missile version of the AGM-84 Harpoon.
U = Underwater. Applies to vehicles designed to attack submarines and underwater targets. Example is the RUM-139 VL-Asroc missile.
W = Weather. Applies to vehicles designed to obtain weather data and collect aerial samples.

Vehicle Type Symbol

For rockets and missiles, the vehicle type symbol identifies the basic vehicle type and will be the final symbol in the mission part of the MDS.

B = Booster. Boosters are primary or auxiliary propulsion units for other vehicles.
M = Missile. Guided missiles are unmanned vehicles flying a path controlled by a guidance system.
N = Probe. Probes are non-orbital unmanned vehicles designed primarily to collect data within the aerospace environment.
R = Rocket. Rockets are single-use unmanned vehicles without guidance after launch.

Exceptions

Of course, there are exceptions – the ASM-135 ASAT is not an Air-Launched Space Support Missile, AIM-135 would have been more appropriate. Convair PQM-102 should have been QF-102 remaining in the aircraft series, as did later aircraft-to-drone conversions (it should also have been designated GQM-102, not PQM-102). Atlas and Titan should be MGM-16 and -25 respectively as they were launched from multiple environments.

There are multiple cases where drones are placed in the missile category; this may have been at the time the most likely place to categorize this type of vehicle. There is now the "Q" series for UAV (Unmanned Aerial Vehicle) craft, and most drones would have been in that category had it existed when the MDS series was inaugurated in 1962. Indeed, the BQM-155 Hunter is now designated the RQ-5. A survey will show that exceptions are seemingly more than the rule for missile designations.

Missile Names

For the most part, names of missiles are assigned during development by the manufacturer, not by the military service or the Department of Defense. They are however approved by the DoD and become part of the accepted nomenclature. These names include "Sparrow," "Javelin" and "Bomarc."

Amendments

While every effort has been made to keep this listing current, accurate and complete, we recognize that errors and omissions are unavoidable. We welcome all comments which will help us improve future editions, or increase the holdings and understanding of the NASM. Please send any documented corrections, additions, comments or images to us at:

National Air and Space Museum
Archives Division
Room 3100, MRC 322
PO Box 37012
Washington, DC 20013-7012
Or e-mail us at nicklasb@si.edu

Acknowledgements

Virtually everyone with the Archives Division of the National Air and Space Museum has assisted with this project; Dana Bell, Kristine Kaske-Martin, Patti Williams, Kate Igoe, Mark Taylor, David Schwartz, Larry Wilson, Paul Silbermann, Mark Kahn, Melissa Keiser, Barbara Weitbrecht and Jessamyn Lloyd. My immediate supervisor Dan Hagedorn (now at the Museum of Flight in Seattle) was extremely supportive of this project and lent advice, support and information when he could, not to mention constant shots of morale boost when he saw my energies flag. Also thank you to Marilyn Graskowiak, the Chair of the NASM Archives Division. Thanks also to the staff of the Department of Space History, especially Frank Winter, and the staff of the Aeronautics Department, especially Bob van der Linden, Roger Connor and Russ Lee. Thank you to the staff of the NASM Branch of the SI Libraries, especially Phil Edwards and Leah Smith. Trish Graboske as our Director of Publications did a wonderful job of displaying patience with me as I tried to give birth to my first book.

Outside of the Museum, I was offered assistance by Peter Alway, Jim Banke, John Bolthouse and Alan Renga of the San Diego Aerospace Museum, Alan D. Fischer and Sara Hammond of Raytheon, John Getsy, Dr. Joe Handelman, Dr. Richard Hallion, Jennings Heilig, Patrick McCarthy, Richard Maltz, Peter W. Merlin, Wes Oleszewski, Jim Rotramel (very special thanks), Kevin M. Rusnak of the Air Force Research Lab History Office, Nancy Rustemeyer, Brandon Silverstein of L-3 BAi, and Matt Steele of ATK, formerly with Orbital Sciences and Steve Zaloga (both a special thanks).

I'd also like to thank the corporations and corporate and military photographers for taking and providing images for this publication, in some cases to Herb Desind, in other cases directly to the Museum to support this publication and other Museum efforts.

I must thank Lionel Leventhal for asking me if I had any ideas during one of his visits to Washington, and his immediately turning to Trish Graboske to approve this. Michael Leventhal took the book along to Frontline Books and has also been a keen supporter of this project, and provided insight and patience. A very special thanks to Deborah Hercun and my editors for massaging and editing this into a quality publication.

Finally a thank you to my parents John and Olean and my four brothers; Jeffery, Martin, Stephen and John, who support me for reasons only they know and understand.

About the Author

Brian Nicklas is a Museum Specialist in the field of aeronautics at the Smithsonian National Air and Space Museum. He has been with the museum for over 20 years and he is also a freelance writer and aviation photographer. A Contributing Editor to Air & Space/Smithsonian magazine, he has been published in magazines and contributed to numerous publications. In his leisure time, Nicklas works with antique aircraft, classic automobiles and models of aircraft and rockets. An alumnus of the Aeronautical Studies program at Embry-Riddle Aeronautical University, he lives in Bethesda, Maryland.

Numerical List of Missiles

Martin **MGM-1** Matador
General Dynamics (Convair) **RIM-2** Terrier
Western Electric **MIM-3** Nike Ajax
Hughes **AIM-4** Falcon
JPL/Firestone **MGM-5** Corporal
Vought **RGM-6** Regulus
Raytheon **AIM/RIM-7** Sparrow
Bendix **RIM-8** Talos – **MQM-8** Vandal
Raytheon (Philco/G.E.) **AIM-9** Sidewinder
Boeing **CIM-10** Bomarc
Chrysler **PGM-11** Redstone
Martin **AGM-12** Bullpup
Martin **MGM/CGM-13** Mace
Western Electric **MIM-14** Nike Hercules
Vought **RGM-15** Regulus II
General Dynamics (Convair) **CGM/HGM-16** Atlas
Douglas **PGM-17** Thor
Martin **MGM-18** Lacrosse
Chrysler **PGM-19** Jupiter
McDonnell **ADM-20** Quail
Nord **MGM-21**
Aérospatiale (Nord) **AGM-22**
Raytheon **MIM-23** Hawk
General Dynamics (Convair) **RIM-24** Tartar
Martin **HGM/LGM-25** Titan
Hughes **AIM-26** Falcon
Lockheed **UGM-27** Polaris
North American **AGM-28** Hound Dog
JPL/Sperry **MGM-29** Sergeant
Boeing **LGM-30** Minuteman
Martin Marietta **MGM-31** Pershing
Aérospatiale (Nord) **MGM-32** Entac
Northrop (Radioplane) **MQM-33**
Teledyne Ryan **AQM/BQM/MQM/BGM-34** Firebee
Northrop (Radioplane) **AQM-35**
Northrop (Radioplane) **MQM-36** Shelduck
Beech **AQM-37**
Northrop (Radioplane) **AQM-38**
Beech **MQM-39**
Globe **MQM-40** Firefly
Fairchild **AQM-41** Petrel
North American **MQM-42** Redhead/Roadrunner
General Dynamics **FIM-43** Redeye
Goodyear **UUM-44** Subroc
Texas Instruments **AGM-45** Shrike
General Dynamics **MIM-46** Mauler
Hughes **AIM-47** Falcon
Douglas **AGM-48** Skybolt
Western Electric/McDonnell-Douglas **LIM-49** Nike Zeus/Spartan
Bendix **RIM-50** Typhon LR
Ford **MGM-51** Shillelagh
LTV **MGM-52** Lance
Rockwell **AGM-53** Condor
Raytheon (Hughes) **AIM-54** Phoenix
Bendix **RIM-55** Typhon MR
Nord/Bell **PQM-56**
Northrop (Radioplane) **MQM-57** Falconer
Aerojet General **MQM-58** Overseer
(USN) APL **RGM-59** Taurus (unbuilt)
Lockheed **AQM-60** Kingfisher
Beech **MQM-61** Cardinal
Martin Marietta **AGM-62** Walleye
USN **AGM-63** (unbuilt)
Rockwell (North American) **AGM-64** Hornet
Raytheon (Hughes) **AGM-65** Maverick
Raytheon (General Dynamics) **RIM-66** Standard MR
Raytheon (General Dynamics) **RIM-67** Standard ER
Air Force Weapons Lab **AIM-68**
Boeing **AGM-69** SRAM
M-70
Raytheon (Hughes) **BGM-71** TOW
Ford **MIM-72** Chaparral
Lockheed **UGM-73** Poseidon
Northrop **MQM/BQM-74** Chukar
USAF **BGM-75** AICBM (unbuilt)
Hughes **AGM-76** Falcon
McDonnell-Douglas **FGM-77** Dragon

General Dynamics **AGM-78** Standard ARM
Martin Marietta **AGM-79** Blue Eye
Chrysler **AGM-80** Viper
Teledyne Ryan **AQM-81** Firebolt
USAF **AIM-82** (unbuilt)
Texas Instruments **AGM-83** Bulldog
Boeing (McDonnell-Douglas)
   AGM/RGM/UGM-84 Harpoon
USN **RIM-85** (unbuilt)
Boeing **AGM-86** ALCM
General Electric **AGM-87** Focus
Raytheon (Texas Instruments) **AGM-88** HARM
UGM-89 Perseus / STAM
USN **BQM-90** (unbuilt)
Teledyne Ryan **AQM-91** Firefly
Raytheon (General Dynamics) **FIM-92** Stinger
E-Systems **GQM-93**
Boeing **GQM-94** B-Gull
Hughes **AIM-95** Agile
Lockheed **UGM-96** Trident I
General Dynamics **AIM-97** Seekbat
Teledyne Ryan **GQM-98** R-Tern
USA **LIM-99** (unbuilt)
USA **LIM-100** (unbuilt)
USN **RIM-101** (unbuilt)
General Dynamics/Sperry **PQM-102** Delta Dagger
Teledyne Ryan **AQM-103**
Raytheon **MIM-104** Patriot
Lockheed **MQM-105** Aquila
USAF FDL **BQM-106** Teleplane
Raytheon (Beech) **MQM-107** Streaker
NWC **BQM-108**
Raytheon (General Dynamics)
   BGM/RGM/UGM-109 Tomahawk
LTV **BGM-110**
Teledyne Ryan **BQM-111** Firebrand
Rockwell **AGM-112**
NSWC **RIM-113** Anti-Cruise Missile Weapon (unbuilt)
Boeing/Lockheed Martin (Rockwell/Martin Marietta) **AGM-114** Hellfire
Euromissile/Hughes/Boeing **MIM-115** Roland
Raytheon (General Dynamics) **RIM-116** RAM

RS Systems **FQM-117** ARCMAT
Martin Marietta **LGM-118** Peacekeeper
Kongsberg **AGM-119** Penguin
Raytheon (Hughes) **AIM-120** AMRAAM
Boeing **CQM/CGM-121** Pave Tiger/Seek Spinner
Motorola **AGM-122** Sidearm
Emerson Electric **AGM-123** Skipper II
Hughes **AGM-124** Wasp
Boeing **RUM/UUM-125** Sea Lance
Beech **BQM-126**
Martin Marietta **AQM-127** SLAT
USN **AQM-128** Drone (unbuilt)
Raytheon (General Dynamics) **AGM-129** ACM
Boeing (Rockwell) **AGM-130**
Boeing **AGM-131** SRAM II
MBDA (BAe Dynamics/Matra) **AIM-132** ASRAAM
Lockheed Martin **UGM-133** Trident II
Martin Marietta **MGM-134** Midgetman
Vought **ASM-135** ASAT
Northrop **AGM/BGM-136** Tacit Rainbow
Northrop **AGM/MGM-137** TSSAM
Boeing **CEM-138** Pave Cricket
Lockheed Martin (Loral) **RUM-139** VL-Asroc
Lockheed Martin (LTV) **MGM-140** ATACMS
IMI (Brunswick) **ADM-141** TALD
Rafael/Lockheed Martin **AGM-142** Have Nap
USAMICOM **MQM-143** RPVT
DoD **ADM-144** (unbuilt)
Teledyne Ryan **BQM-145** Peregrine
Oerlikon/Lockheed Martin **MIM-146** ADATS
BAI Aerosystems **BQM-147** Exdrone
Raytheon/Lockheed Martin **FGM-148** Javelin
PQM-149 UAV-SR / McDonnell-Douglas Sky Owl
PQM-150 UAV-SR
AeroVironment **FQM-151** Pointer
AIM-152 AAAM
USAF **AGM-153** (unbuilt)
Raytheon (Texas Instruments) **AGM-154** JSOW
TRW/IAI **BQM-155** Hunter
Raytheon **RIM-156** Standard SM-2ER Block IV
Raytheon **MGM-157** EFOGM

Lockheed Martin **AGM-158** JASSM
Boeing (McDonnell-Douglas) **AGM-159** JASSM
Northrop Grumman (Teledyne Ryan) **ADM-160** MALD
Raytheon **RIM-161** Standard SM-3
Raytheon **RIM-162** ESSM
Orbital Sciences **GQM-163** Coyote
Lockheed Martin **MGM-164** ATACMS II
Raytheon **RGM-165** LASM
Lockheed Martin **MGM-166** LOSAT/KEM
Composite Engineering **BQM-167** Skeeter
Lockheed Martin **MGM-168** ATACMS Block IVA
Lockheed Martin **AGM-169** JCM Joint Common Missile
Griffon Aerospace **MQM-170** Outlaw UAV
Griffon Aerospace **MQM-171** BroadSword UAV
Lockheed-Martin **FGM-172** Predator / SRAW
Alliant Techsystems **MQM-173** MSST
Raytheon **RIM-174** ERAM

American Missiles

MGM-1 Matador

Right ½ front view of a MGM-1 (B-61) Matador Missile just at launch from a mobile launcher from the Air Force Missile Test Center, Cocoa, Florida. September 20, 1951.

Right ½ rear view of a MGM-1 Matador missile on launch platform, showing the detail of the Rocket Assisted Take-off (RATO) motor under the rear fuselage. Appears to be a test round at launch from Holloman AFB, New Mexico early in the Matador program.

Specifications

Length: 39 ft 8 in (1209.04 cm)
Diameter: 4 ft 6 in (137.16 cm)
Maximum Span: 27 ft 11 in (850.90 cm)
Height: 9 ft 8 in (294.64 cm)
Speed: over 650 mph
Function: Surface-to-Surface
Weight: 13,800 lbs (6,265 kg)
Warhead: W5 Nuclear Warhead, 20 KT
Guidance: Ground Guidance
First Use Date: 1949 – Holloman AFB, NM
Producer: Martin Company, Baltimore, Maryland
Users: US Air Force, Germany
Other Designations: B-61, TM-61, T-50
Status: Was operational, withdrawn from use

The Matador has been described as the first operational guided weapon used by the US Air Force as all preceding weapons were unguided, the distinction between unguided rockets and missiles like the MGM-1. (A rocket can be part of a missile, but not vice-versa.) The Matador was originally designated as a bomber, the B-61 before being used operationally as the TM-61, could be equipped with a conventional or nuclear warhead. The Matador could be distinguished from the latter similar looking missile the Mace by the Matador's shorter, yet more pointed nose cone. The Matador could be fired from mobile launchers or from fixed emplacements. Over 1,200 Matadors were produced by Martin, and the last of the operational missiles were replaced in 1962. After which the MGM-1 designation was applied.

RIM-2 Terrier

Right ½ front view of shipboard launcher on the USS *Mississippi* (EAG-128) holding two RIM-2 Terrier missiles.

Left ¼ front view of shipboard launcher on the USS *Canberra* (CAG-2) holding two RIM-2 Terrier missiles.

Specifications

Length: 27 ft (823 cm)
Diameter: 1 ft 2 in (35.56 cm)
Maximum Span: 4 ft (121.92 cm)
Function: Surface-to-Air
Weight: 3,300 lbs (1,498 kg)
Warhead: 220 lbs, High Explosive
Guidance: Radar Guided
First Use Date: 1951
Producer: Convair
Users: US Navy, Italy, Netherlands
Other Designations: SAM-N-7
Status: Was operational, withdrawn from use

The Terrier was the first operational surface-to-air missile used by the US Navy. Guided missile cruisers were equipped with multiple launchers, each launcher having two rails each for the RIM-2. The two stage Terrier was initially lofted by a 12 ft long booster, which combined with the main sustainer motor (both solid fuel rocket type) would push the RIM-2 to speeds approaching Mach 2.5. The Terrier was equipped with a conventional explosive warhead.

MIM-3 Nike-Ajax

Left ¾ front view of Site 40, Battery C, 865th AAA Missile Battalion, one of 16 units of the 108th Arty Group (AD). The site is located at Long Beach, California and is equipped with MIM-3 Nike-Ajax missiles.

View of an erected MIM-3 Nike-Ajax missile at White Sands Missile Range, New Mexico.

Specifications

Length: 34 ft 10 in (1061.72 cm)
Diameter: 12 in (30.48 cm)
Maximum Span: 4 ft 6 in (137.16 cm)
Function: Surface-to-Air
Weight: 2,259 lbs (1,025 kg)
Warhead: (3) High Explosive, Fragmentation
Guidance: Radar Commanded
First use: 1952
Producer: Western Electric with Douglas Aircraft
Users: US Army
Other Designations: SAM-A-7
Status: Was operational, withdrawn from use

The first anti-aircraft surface-to-air missile used by the United States military, the MIM-3 Nike Ajax would come to symbolize air defense of American cities for years to come, by virtue of the Nike name. Development began post-World War Two, from ideas formulated during the war when defending cities by the new missile technology became apparent. Cities would be ringed by launch batteries, coordinated through the SAGE (Semi Automatic Ground Environment) command and control network. The conventionally-armed Nike Ajax used a solid rocket motor booster, with a difficult to use red fuming nitric acid and kerosene powered sustainer. Nike Ajax was superseded by the MIM-14 Nike Hercules.

AIM-4 Falcon

Left side view of two AIM-4 Falcon missiles. The foreground is the radar-guided AIM-4 previously designated the GAR-3A (later the AIM-4F), and in the back is the infrared-guided Falcon previously called the GAR-4A (later the AIM-4G).

Left side view of two AIM-4 Falcon missiles. On the left is the infrared-guided AIM-4 previously designated the GAR-2A (later the AIM-4C), and on the right is the radar-guided Falcon previously called the GAR-1D (later the AIM-4A). The aircraft in the background is the Convair F-102A Delta Dagger.

Specifications (AIM-4F)

Length: 7 ft (213.36 cm)
Diameter: 6½ in (184.15 cm)
Maximum Span: 2 ft (60.96 cm)
Function: Air-to-Air
Weight: 150 lbs (68 kg)
Warhead: 7.5 lbs, High Explosive
Guidance: Semi-active Radar, Infrared Seeker
First Use or Design Date: 1960
Producer: Hughes
Users: US Air Force, Canada, Sweden, Finland, Japan, Switzerland. Greece, Taiwan, Turkey and Japan.
Other Designations: AAM-A-2, XF-98, GAR-1, GAR-2, GAR-3, GAR-4, HM-58, HM-68, Rb 28
Status: Was operational, withdrawn from use

The AIM-4 was the first guided weapon to enter service with USAF fighter units. Initially used on the F-89 Scorpion, it was later used on the F-101 Voodoo, F-102 Delta Dagger and F-4 Phantom II. It was phased out of service with the last aircraft to carry it, the F-106 Delta Dart, when it was withdrawn from operational use in 1988. The Falcon used several guidance systems, depending on the variant used, either an infrared seeker or radar guidance. The AIM-4 scored five victories in Southeast Asia against North Vietnamese MiGs, but it was not considered a successful missile in Vietnam. The Falcon name was used on other missiles that were a continuation of the design evolution, such as the AGM-76.

MGM-5 Corporal

FORT BLISS, TEXAS – Left side view of MGM-5 Corporal missile on M-2 self propelled erector. Photo taken in the 2300 training area.

View of launch area as an MGM-5 Corporal missile is erected on M27 launcher. Support truck in foreground, cherry-picker truck for missile maintenance in background.

Specifications

Length: 46 ft (1402 cm)
Diameter: 2 ft 6 in (76.2 cm)
Maximum Span: 7 ft (213.36 cm)
Function: Surface-to-Surface
Weight: 12,000 (5,448 kg)
Warhead: High Explosive or W7 Nuclear Warhead, 20 KT
Guidance: Command Guided
First Use Date: 1954
Producer: Jet Propulsion Lab
Users: US Army, United Kingdom
Other Designations: SSM-A-17, M-2
Status: Was operational, withdrawn from use

The Corporal was a short range surface-to-surface missile with a nuclear capability. The MGM-5 was built by Firestone (airframe) and Ryan (engine) but these firms were subcontracted to the Jet Propulsion Lab (JPL) of the California Institute of Technology as the prime contractor. The Corporal was America's first operational guided ballistic missile, even though it was shorter ranging than ballistic missiles that would follow.

The MGM-5 was the first American missile able to loft a kiloton nuclear warhead, although the Corporal was also planned to use a conventional warhead. Used as a tactical missile, the choice of warhead was dictated by the needs of the battle at the time. Although replaced by the MGM-29 Sergeant missile, it remained in service until 1966, at that point serving solely with the British Army.

RGM-6 Regulus I

Right ½ front view of US Navy submarine USS *Greyback* (SSG-574) entering San Diego harbor with RGM-6 Regulus missile amidships.

Left ½ rear view from above of a RGM-6 (previously SSM-N-8) Regulus missile on a firing platform on the deck of the aircraft carrier USS *Hancock* (CVA-19) during PROJECT RAM off San Diego, California. February 7, 1955.

Specifications

Length: 33 ft (1005.84 cm)
Diameter: 4 ft 6 in (137.16 cm)
Maximum Span: 21 ft (640.08 cm)
Function: Surface-to-Surface
Weight: 14,522 lbs (6,593 kg)
Warhead: W5 Nuclear Warhead, 20 KT, W27 Nuclear Warhead, 2 MT
Guidance: Command Guided
First Use Date: 1951
Producer: Chance Vought
Users: US Navy
Other Designations: SSM-N-8, KDU-1
Status: Was operational, withdrawn from use

The Regulus I was a surface-to-surface guided missile designed for submarine launch, but was also carried by US Navy aircraft carriers and several cruisers. Powered by an air breathing turbojet engine, the Regulus I was a basic entry to bring the US Navy into the missile age before the operational capabilities of missiles capable of being fired from submerged submarines became a reality. The RGM-6 was used for almost ten years in the US Navy. The design resembled a small fighter aircraft, minus the cockpit. To assist in the launch, the Regulus was provided with two solid rocket boosters which dropped away shortly after launch. A recoverable variant was developed for training sailors in its operation.

AIM-7 Sparrow / RIM-7 Sea Sparrow

Left side view of McDonnell-Douglas F-15A Eagle fighter as it launches an AIM-7 Sparrow missile. The F-15A is assigned to 85th Test Squadron, Eglin AFB, Florida.

Right ½ front view a AIM-7 Sparrow missile mounted on a Kaman UH-2C Seasprite helicopter parked at Naval Air Station Point Mugu, California. Photo dated July 1972.

Specifications

Length: 12 Feet (365.76 cm)
Diameter: 8 in (20.32 cm)
Maximum Span: 3 ft 4 in (101.6 cm)
Function: Air-to-Air
Weight: 500 lbs (227 kg)
Warhead: 66-88 lbs, High Explosive, Fragmentation
Guidance: Radar
First Use Date: 1966
Producer: Raytheon
Users: United States, Canada, Japan, Germany, Iran, Israel, Italy, Pakistan, Saudi Arabia, South Korea, Thailand, United Kingdom
Other Designations: AAM-N-2 (AIM-7A Sparrow I), AAM-N-3 (AIM-7B Sparrow II), AAM-N-6 (AIM-7C Sparrow III)
Status: Operational

The AIM-7 is a radar-guided air-to-air missile, used by most US military fighter aircraft from the 1950s to the present. There is a navalized variant for surface-to-air use designated the RIM-7. The Sparrow is carried aboard aircraft on weapons pylons, while Sea Sparrow is launched from enclosed containers on deck mounts on surface ships. As a radar-guided weapon, it was the primary beyond visual range missile on American fighters of all branches of service until recently, when it has been replaced by the AIM-120. This after many missiles were proposed and abandoned as Sparrow replacements.

RIM-8 Talos

Right ½ front view of a RIM-8 Talos missile mounted on its launcher aboard the guided missile light cruiser USS *Galveston* (CLG-3). Photo dated February 1959.

Close right ½ front view of a RIM-8 Talos missile mounted on its launcher aboard the guided missile light cruiser USS *Galveston* (CLG-3). Note the seaman standing in the hatch on the central mount of the launcher. Photo dated May 28, 1959.

Specifications

Length: 31 ft 3 in (952.5 cm)
Diameter: 2 ft 6 in (72.2 cm)
Maximum Span: 9 ft 6 in (289.56 cm)
Function: Surface-to-Air
Weight: 7,000 lb (3,178 kg)
Warhead: 300 lbs, High Explosive, W30 Nuclear Warhead, 2-5 KT
Guidance: Beamriding, Semi-active Homing
First Use Date: 1959
Producer: Bendix Corporation
Users: US Navy
Derivatives: MQM-8 Vandal
Other Designations: SAM-N-6
Status: Was operational, withdrawn from use

Developed by the Applied Physics Lab of Johns Hopkins University, Talos was built with the Bendix Corporation as the prime contractor. A radar-guided surface-to-air weapon, the RIM-8 "rode a beam" of targeting radar to interception. The Talos, which could carry either conventional or nuclear warheads, could also be directed at surface targets. The Talos ended its career with the Navy as a target drone to simulate enemy anti-ship missiles for training with the next generation of ship to air missiles. This expendable target is known as the MQM-8 Vandal.

AIM-9 Sidewinder

The newest-generation Sidewinder, the AIM-9X.

Right side view of a Northrop F-20 Tigershark with AIM-9 Sidewinder missiles mounted on the wingtip launch rails.

Aviation Ordnanceman 2nd Class Daniel Porter checks an AIM-9 Sidewinder missile for proper mounting on an F/A-18C Hornet, from the "Blue Diamonds" of Strike Fighter Squadron (VFA) 146, aboard *Nimitz*-class aircraft carrier USS *John C. Stennis* (CVN 74) prior to flight operations in 2007. Note the moveable double delta forward fins and the rollerons on the rear fin tips, signs of the mid-generation Sidewinder.

Pilot stands next to an AIM-9 Sidewinder missile in front of a Lockheed F-104A Starfighter. The pilot wears an MC-3 partial pressure suit and MA-2 helmet for high-altitude flight.

Specifications

Length: 9 ft 5 in (287.02 cm)
Diameter: 5 in (12.7 cm)
Maximum Span: 2 ft ¾ in (62.86 cm)
Function: Air-to-Air
Weight: 200 lbs (90 kg)
Warhead: 20 lbs, High Explosive, Fragmentation
Guidance: Infrared
First Use or Design Date: 1952
Producer: Philco (Ford Aerospace), General Electric, Mitsubishi, Hughes, Loral, Raytheon
Users: United States, Great Britain, Canada, Sweden, Japan, Germany, 30+ more
Similar Types: Atoll (Soviet Union), Shafir (Israel), Kukri (South Africa)
Other Designations: AAM-N-7, GAR-8
Status: Operational

The most successful of the short-range air-to-air missiles, the AIM-9 Sidewinder is used around the world, including the non-licensed near copies used by some nations.

First developed by the Naval Ordnance Test Station at China Lake, California in the early 1950s, the Sidewinder was originally designated the AAM-N-7. Production of the missile was done by Philco, the electronics arm of Ford Motor Company (later Ford Aerospace). The first AIM-9 successfully fired was in 1953.

While used very successfully by all operators of the missile, it was Taiwan that recorded the first success with the Sidewinder, when Nationalist Chinese F-86F Sabres shot down four Peoples Republic of China MiG-17s on September 24, 1958.

Constant refinements in the Sidewinder have given the missile a changing appearance from variant to variant, from the nose contours changing to the fin shape, size and location in an ever evolving program of design.

The Sidewinder has been flown in many variants, but these can be divided into at least three generations of the missile. The first generation was limited to attacks from the rear of the target so that the exhaust was a clear heat source for the seeker head, the mid generation was able to engage from all-aspects (including head-on attacks), and the newest generation, exemplified by the AIM-9X has a 90 degree off-boresight capability, so the Sidewinder no longer needs to be pointed directly at the target. The AIM-9X also has a redesigned missile body: fixed forward fins, moveable rear fins, with vectored rocket nozzles and other improvements removing the need for the rear fins to incorporate the gyroscopic rollerons, otherwise a distinctive feature of the Sidewinder and its clones.

During the 1960s to 1990s, it was common to see the AIM-9 Sidewinder as a short range missile teamed up on the same fighter together with its stable mate, the AIM-7 Sparrow, a medium range radar-guided missile. The size and shape of the Sidewinder have made it easy to carry on smaller fighter aircraft, and it has, in fact, become a design element in some aircraft.

CIM-10 /CQM-10 Bomarc

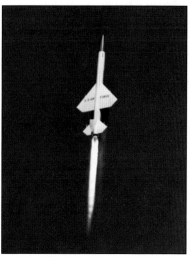

View of the above-ground launch coffins for the CIM-10 Bomarc missile. Four Bomarc missiles are raised to the vertical launch position.

View of a CIM-10 Bomarc missile in flight shortly after launch from Cape Canaveral, Florida.

Specifications (CIM-10B)

Length: 45 ft 1 in (1374.13 cm)
Diameter: 2 ft 11 in (88.9 cm)
Maximum Span: 18 ft 2 in (553.72 cm)
Function: Surface-to-Air
Weight: 16,032 lb (7,278 kg)
Warhead: 1,000 lbs, High Explosive or W40 Nuclear Warhead 7-10 KT
Guidance: Command guided to target, active homing during final phase
First Use Date: 1957
Producer: The Boeing Company
Users: US Air Force, Canada
Other Designations: XF-99, IM-99
Status: Was operational, withdrawn from use

The underslung ramjet engines are a signature to the look of the CIM-10 Bomarc. The fuselage housed the initial means of flight: on the CIM-10A, it was a liquid-fueled engine, on the CIM-10B it was a more stable solid-fuel engine. Once at altitude, the Bomarc's liquid fuel ramjet engines would take over, to deliver the missile to its target.

The instability of the liquid fuel was dramatically demonstrated when a Bomarc A at McGuire AFB, New Jersey, exploded in June 1960. The subsequent fire destroyed the warhead casing, which was a nuclear warhead, and there was a subsequent release of plutonium in the immediate area. Bomarc A bases outnumbered the need for the Bomarc B, so two bases were shutdown when the transition was made. But by 1973, all Bomarc missile sites, including two operated by the Canadians, were shut down. Starting in 1962, some Bomarcs were modified to be used as high-altitude supersonic aerial targets as the CQM-10.

PGM-11 Redstone

Side view of the PGM-11 Redstone Missile System lightweight erection equipment being demonstrated on the outside hardstand and being picked up by remote control for transmission to the classroom through Closed Circuit Television. The missile is braced and secured to the launch platform as cables pull it upright. Note the PGM-19 Jupiter missile in the background.

Side view of a PGM-11 Redstone missile test round on launch platform at White Sands Missile Range, New Mexico.

Specifications

Length: 69 ft 4 in (2113.27 cm)
Diameter: 5 ft 10 in (177.8 cm)
Maximum Span: 12 ft 9 in (388.62 cm)
Function: Surface-to-Surface
Weight: 61,000 lbs (27,694 kg)
Warhead: Nuclear Warhead 1 – 3.75 MT
Guidance: Inertial
First Use Date: 1953
Producer: Chrysler
Users: US Army
Other Designations: SSM-A-14
Status: Was operational, withdrawn from use

A simple view of the Redstone missile was that it was an Americanized improvement of Germany's WWII V-2 rocket. As the naming convention for the Army at the time was to name missiles and rockets after the military rank structure, the PGM-11 (then the SSM-A-14) was known as the Major missile. This must have led to confusion with the name questioned as "What major missile?" with the reply "The one from the Redstone Arsenal." So Redstone became the official name after unofficial use. The Redstone was launched then discarded the booster, leaving a guidance section with the conventional or nuclear warhead to continue to the target. Redstone later rose to fame as the booster for the first two manned Mercury flights, the sub-orbital ventures of Alan Shepard and Gus Grissom.

AGM-12 Bullpup

Right ½ front view of a AGM-12 Bullpup missile mounted on a North American F-100 Super Sabre fighter.

Rear view from above of a Douglas A-4 (A4D) Skyhawk just after it has launched an AGM-12 Bullpup missile at a ground target.

Specifications (AGM-12C)

Length: 10 ft 6 in (320.04 cm)
Diameter: 1 ft (30.98 cm)
Maximum Span: 3 ft 1 inch (93.98 cm)
Function: Air-to-Surface
Weight: 1,784 lbs (810 kg)
Warhead: 250–1000 lbs, High Explosive or W45 Nuclear Warhead
Guidance: Radio Command
First Use Date: 1959
Producer: Martin
Users: US Navy, US Marine Corps, US Air Force, Australia, Denmark, Greece, Israel, Norway, Taiwan, Turkey, United Kingdom
Other Designations: GAM-83/ ASM-N-7A (AGM-12B Bullpup A 250lb warhead), ASM-N-7B (AGM-12C Bullpup B 1000lb warhead)
Status: Was operational, withdrawn from use

The Bullpup was an air-to-surface missile with the capability of striking targets with a conventional or nuclear warhead, depending on the sub-variant of the AGM-12 in use. The Bullpup could be carried by a variety of patrol and attack aircraft, from the A-4 Skyhawk to the P-3 Orion. The AGM-12 also armed the Buccaneer of the Royal Navy's Fleet Air Arm. Many Bullpups were built, but few were used as the Bullpup was not "user friendly," the guidance required the pilot to guide the missile with a joystick in the cockpit, while also flying his aircraft! A simplified training version was based on the 5-in rocket, with an added guidance package; this was much smaller than the operation Bullpup.

CGM-13 Mace

Left side view of an MGM-13 Mace missile just after launch. The RATO motor used to assist the missile to flight speed is attached to the rear of the missile and is still burning.

Left ½ front view of an MGM-13 (previously the TM-76) Mace missile on its launcher. Overhead a Boeing KC-135 Stratotanker refuels a Boeing B-52 Stratofortress.

Specifications

Length: 44 ft (1341.12 cm)
Diameter: 4 ft 6 in (137.16 cm)
Maximum Span: 22 ft 11 in (698.5 cm)
Function: Surface-to-Surface Missile
Weight: 18,000 lbs (8,172 kg)
Warhead: High Explosive or Nuclear Warhead
Guidance: Terrain Matching (Mace A) Inertial (Mace B)
First Use Date: 1956
Producer: Martin
Users: US Air Force
Other Designations: TM-61B, TM-76
Status: Was operational, withdrawn from use

The CGM-13 Mace was developed from the MGM-1 Matador, and initially shared the designation used under the previous designation system as the TM-61, the Mace as the TM-61B. As differences between the two missiles grew, it was redesignated the TM-76, until the current system applied the CGM-13 designation. The Mace had a more rounded warhead and like Matador could be conventional or nuclear armed. Mace was more easily transported, having shorter wings that folded, as opposed to the wing of the Matador that required removal for the missile to be moved on its transporter/erector. The Mace was designated as the CGM-13 for those launched from hardened shelters, and MGM-13 for those fired from mobile launchers.

MIM-14 Nike-Hercules

Left side view of two MIM-14 Nike-Hercules missiles as they are raised into launch position.

View of an MIM-14 Nike-Hercules guided missile as it zooms skyward a fraction of a second after blast-off at the Ft. Bliss, Texas McGregor Range.

Specifications

Length: 41 ft 6 in (1264.92 cm)
Diameter: 2 ft 1 inch (63.5 cm)
Maximum Span: 6 ft 2 in (187.96 cm)
Function: Surface-to-Air
Weight: 5,000 lbs (2,270 kg)
Warhead: T45 1,100 lbs, High Explosive or W31 Nuclear Warhead, 2-40 KT
Guidance: Radar Command
First Use Date: 1955
Producer: Western Electric
Users: US Army, Belgium, Denmark, Germany, Greece, Italy, Japan, Republic of Korea, Netherlands, Norway, Turkey
Other Designations: SAM-A-25
Status: Was operational, withdrawn from use

Nike-Hercules could be considered a major improvement on the MIM-3 Nike-Ajax in that it used much the same launch system and support equipment, but with increased reliability and safety with advances of technology and manufacturing. The MIM-14 also used the same booster as Nike-Ajax, albeit with four MIM-3 boosters strapped together for use on the Nike-Hercules. The solid fuel was also used in the Nike-Hercules sustainer. The performance of this improved weapon allowed for the use of conventional as well as nuclear warheads. Enough Nike missiles had been produced of both variants (MIM-3 and MIM-14) to allow for the use after withdrawal from defense duty to supply Nike sounding rocket motors to the scientific community for years to come.

RGM-15 Regulus II

A RGM-15 (SSM-N-9) Regulus II missile is launched from the deck of the USS *King County* (AG-157) during trials of the missile to test compatibility with submarines. A faux submarine deck was built upon the *King County* for these trials.

Right ½ front view of a RGM-15 (SSM-N-9) Regulus II missile shortly after launch at Edwards AFB, California.

Specifications

Length: 67 ft 2 in (2047.24 cm)
Diameter: 6 ft (182.88 cm)
Height: 12 ft 6 in (381 cm)
Maximum Span: 20 ft ½ inch (610.87 cm)
Function: Surface-to-Surface
Weight: 22,000 lbs (9,988 kg)(minus booster)
Warhead: W27 Nuclear Warhead, 2 MT
Guidance: Inertial
First Use Date: May 29, 1956
Producer: Chance Vought
Users: US Navy
Other Designations: SSM-N-9
Status: Tested only

The SSM-N-9 (RGM-15) was to have replaced the RGM-6 in fleet service with the Navy, but the advent of the Polaris missile ended the career of the Regulus II before it could begin. First launched in 1957, the Regulus II was capable of supersonic speeds, a vast improvement over the subsonic Regulus I. The RGM-15 was also planned for land launch, using a specialized trailer. Early versions were built with landing gear for ease of operation and for recovery to accelerate the test process. The first submarine launch of a Regulus II was in September 1958 from the USS *Grayback*, but before the end of that year further work on the Regulus II was cancelled and production halted. The Regulus II was actually re-designated as the RGM-15 long after the program was cancelled.

HGM-16/CGM-16 Atlas

Right side view of a CGM-16 Atlas missile upon its transport trailer at Vandenberg AFB, California.

Side view of an HGM-16 Atlas missile upon its launch platform raised atop the underground silo at Vandenberg AFB, California. Dated 1962.

Specifications (Atlas D)

Length: 82 ft 6 in (2514.6 cm)
Diameter: 10 ft (304.8 cm)
Span (at engines): 16 ft (487.68 cm)
Function: surface-to-surface ICBM
Weight: 267,136 lbs (121,172 kg)
Warhead: W38 Nuclear Warhead, 4 MT (Atlas F)

Guidance: Radio Command, Inertial
First Use Date: 1957
Producer: Convair
Users: US Air Force
Other Designations: SM-65
Status: Was operational, withdrawn from use

The US Air Force's Strategic Air Command's first true ICBM, or Intercontinental Ballistic Missile, was the Atlas. The missile became operational in 1959 and was operated at eight Air Force bases across the United States.

The Atlas was considered by some to be a "stage and half" missile. That is because of the three engines, only one was used throughout the flight. The outer two engines would slip away after the boost phase to conserve launch weight and enhance efficiency. Another aspect of the Atlas was its lightweight construction, using very few internal structures, but rather pressurization to maintain structural rigidity. This "balloon" construction meant that the missile would collapse under its own weight if the tanks were not kept full, either by pressurized air or by fuel.

The CGM-16 Atlas was coffin launched, that is, it was stored on its side, and raised from a shelter into position for fueling and launch. The HGM-16 was stored vertically in a silo, but raised above ground when on alert or for launch.

The highlight of the Atlas program, and the reason for most of its fame was the use of several "man-rated" Atlases to launch the American Mercury spacecraft into orbit, John Glenn being the first astronaut to achieve orbit riding an Atlas.

By 1965, the Atlas was being phased out, the missiles becoming launch vehicles for test programs of satellites, or used as targets for other missiles such as Nike Hercules. Further development of the Atlas continues, but in a strange quirk of fate, many are launched using Russian engines, likely made at factories that had been targeted by the earlier generation of Atlas ICBMs.

PGM-17 Thor

View of the PGM-17 Thor missile standing fueled atop its launch platform next to the storage pad hangar.

View of two PGM-17 Thor missiles standing atop their launch platforms. "Thor missiles, in RAF livery, are shown on station in England after one of the most difficult tasks of its kind ever undertaken. The first Thor wing of 15 missiles was turned over 3½ years after the initial authorization by the US Department of Defense."

Specifications

Length: 65 ft (1981.2 cm)
Diameter: 8 ft (243.84 cm)
Function: Surface-to-Surface
Weight: 110,000 lbs (49,940 kg)
Warhead: W49 Nuclear Warhead, 1MT
Guidance: Inertial
First Use Date: 1958
Producer: Douglas
Users: US Air Force, Italy, Turkey
Other Designations: SM-75, WS-315A)
Status: Was operational, withdrawn from use

The Thor was a single-stage Intermediate Range Ballistic Missile (IRBM) used by both the US Air Force and the Royal Air Force. The size of the PGM-17 was dictated by a requirement that the nuclear-armed missile be transportable by air. In the period of time the Thor was in use, this meant the C-124 and C-133 aircraft. As missile technology rapidly progressed, the Thor was replaced by other missiles. Many Thor missiles were then converted for use as space launch vehicles, either as single stage boosters or adapted to carry a second stage where the warhead had previously been mounted.

MGM-18 Lacrosse

Left side view of an MGM-18 Lacrosse missile atop the 2½-ton transport truck which also serves as the launcher. The soldiers are performing final checks before launch.

Left ½ rear view of an MGM-18 Lacrosse missile as it is launched from the 2½-ton transport truck.

Specifications

Length: 19 ft 2 in (584.2 cm)
Diameter: 1 ft 8 in (50.8 cm)
Maximum Span: 9 ft (274.32 cm)
Function: Surface-to-Surface
Weight: 3,500 lbs (1,589 kg)
Warhead: 250–540 lbs High Explosive or W40 Nuclear Warhead, 1.5-10 KT
Guidance: Radar Guided
First Use Date: 1957
Producer: Martin
Users: US Army, Canada
Other Designations: SSM-A-12
Status: Was operational, withdrawn from use

Developed by the Cornell Aeronautical Laboratory the MGM-18 Lacrosse was a surface-to-surface missile for use against short range targets. A mobile weapon, it was carried and launched from a 2½-ton M45 truck, from a helical rail launcher. The radar guidance of Lacrosse allowed a forward observer to guide the MGM-18 to a target beyond the visual range of the missile. The close support role of the Lacrosse while aiding troops under fire dictated the warhead selected, although the MGM-18 was capable of nuclear as well as conventional explosive warheads.

PGM-19 Jupiter

Specifications

Length: 65 ft 4 in (1991.36 cm)
Diameter: 8 ft 9 in (266.7 cm)
Function: Surface-to-Surface IRBM
Weight: 109,000 lbs (49,487 kg)
Warhead: W49 Nuclear Warhead, 1MT
Guidance: inertial
First use: 1959
Producer: Chrysler
Users: US Air Force
Other Designations: SM-78
Status: Was operational, withdrawn from use

The Jupiter missile was an Intermediate Range Ballistic Missile (IRBM) with a range of approximately 1,500 miles. It was developed by the Army Ballistic Missile Agency (ABMA) for use by the US Army. Initially it was also to be used by the US Navy, but the Navy withdrew, preferring to use solid fuel missiles on submarines, as opposed to dealing with the intricacies of liquid fuels on missiles like Jupiter. After a separation of duties, the Army lost missiles with more than a "battlefield range" and IRBMs such as the PGM-19 became Air Force assets.

The short, squat look of the Jupiter was planned as it was to have served aboard submarines, but this ruled out mobile deployment with the then currently available cargo aircraft, as the PGM-19 was too wide. The Jupiter was deployed by the Air Force overseas after agreements with Italy and Turkey had been reached. Basing the Jupiter in these two countries could allow targeting of the Soviet Bloc given the intermediate range of the missile. Part of the reason for the October 1962 Cuban Missile Crisis was the concern of the Soviet Union over these missiles. Their removal from Turkey and Italy was part of the agreement ending the crisis.

In a bit of naming befuddlement, the PGM-19 Jupiter should not be confused with the Jupiter-C, which was a non-military modified PGM-11 Redstone used for spaceflight activities.

Prelaunch view of a PGM-19 Jupiter missile at the Air Force Missile Test Center, Patrick AFB, Florida.

Launch of a PGM-19 Jupiter missile at the Air Force Missile Test Center, Patrick AFB, Florida – Cape Canaveral Launch Site at 9:07AM (EST), first Jupiter missile fired by an Italian Air Force crew. April 22, 1961.

ADM-20 Quail

Left ½ rear view of an ADM-20 (previously GAM-72) Quail missile on a handling/loading dolly.

Right ½ front view of a Boeing B-52 Stratofortress with a McDonnell ADM-20 Quail Missile in front on a handling dolly. The ADM-20 was launched as a decoy, as it mimicked the radar signature of the B-52. The Quail was initially designated the GAM-72.

Specifications

Length: 12 ft 10 in (391.15 cm)
Height: 3 ft 4 in (101.6 cm)
Maximum Span: 5 ft 4 in (162.56 cm)
Function: Decoy
Weight: 1,200 lb (544 kg)
Guidance: Programmed
First Use Date: 1958
Producer: McDonnell Aircraft
Users: US Air Force
Other Designations: GAM-72
Status: Was operational, withdrawn from use

The Quail was developed to mimic the B-52 Stratofortress bomber, by displaying through shape and electronic countermeasures the radar signature of the much larger aircraft. Launched from the bomb bay of the B-52, the turbo-jet powered ADM-20 would unfold its wings and either continue to the same target area as the B-52, or could be directed to another area entirely. In either case the Quail would distract enemy air defenses by multiplying the perceived threat. Improvements in radar technology soon made the ADM-20 redundant, as it could no longer be confused with the B-52 bomber it was to mimic.

Front view, from above, of an ADM-20 (previously GAM-72) Quail missile on a handling/loading dolly.

MGM-21

Right side view of a US Army sergeant preparing an MGM-21 missile for launch. The MGM-21 was a French design known in France as the SS.10.

Front ¼ left view of an MGM-21 missile in the launcher/shipping container.

Specifications

Length: 2 ft 10 in (86.36 cm)
Diameter: 6 in (15.2 cm)
Maximum Span: 2 ft 5 ½ in (74.93 cm)
Function: Surface-to-Surface
Weight: 33 lbs (15 kg)
Warhead: 11 lbs, High Explosive
Guidance: manual wire guided
First Use Date: 1952
Producer: Nord
Users: US Army, France, Sweden, Switzerland, Germany, Israel
Other Designations: SS.10
Status: Was operational, withdrawn from use

The US Army acquired the SS.10 missile for use by the Seventh Army in Germany. Designated the MGM-21, the missile was guided, as the operator tracked both the target and missile, guiding the missile manually to the target using a wire that uncoiled behind it in flight. This was a cumbersome task, and was replaced by semi-automatic wire guidance, where the operator would only track the target, the missile's electronics guiding it. In French use as the SS.10, the missile was fired from slow aircraft such as helicopters or the Morane-Saulnier MS.505 Criquet, but the US Army limited the MGM-21 to ground use.

AGM-22

Right side view of an AGM-22 missile on a cradle. The AGM-22 was a Nord design known in France as the SS.11.

Left side view of a Bell UH-1 Iroquois helicopter firing one of six AGM-22 missiles off the launch rails.

Specifications

Length: 4 ft (121.92 cm)
Diameter: 6½ in (16.50 cm)
Maximum Span: 1 ft 6 in (45.72 cm)
Function: Air-to-Surface
Weight: 66 lbs (30 kg)
Warhead: 15 lbs, High Explosive Shaped Charge
Guidance: Wire Guided
First Use Date: 1955
Producer: Nord
Users: US Army, France, Abu Dhabi, Argentina, Bahrain, Belgium, Brazil, Brunei, Canada, Denmark, Ethiopia, Finland, Germany, Greece, India, Iran, Iraq, Israel, Italy, Ivory Coast, Lebanon, Libya, Malaysia, Netherlands, Norway, Peru, Portugal, Saudi Arabia, Senegal, South Africa, Spain, Sweden, Switzerland, Tunisia, Turkey, United Kingdom, Uganda, Venezuela
Other Designations: Nord SS.11
Status: Was operational, withdrawn from use

Front view, from below, of a Bell UH-1 Iroquois helicopter with a load of six AGM-22 missiles on launch rails.

Developed as a surface-to-surface missile by France, the SS.11 was acquired by the US Army as an air-to-surface missile to be deployed on helicopters, notably the Bell UH-1 Iroquois (aka Huey). The MGM-22 was wire-guided, and with a skilled operator would fly a path with a plus or minus 3 ft deviation. This was evidently quite a feat to perform from a moving helicopter at a moving armored vehicle in the heat of combat, especially so if the vehicle was shooting back at the helicopter.

MIM-23 Hawk

Left side view of a M501 transporter holding three MIM-23 Hawk missiles. The M501 would transport and load all three missiles at once onto a launcher.

Left ½ front view of a trailer launcher for the MIM-23 Hawk missile. The launcher held three MIM-23 missiles at once.

Specifications

Length: 16 ft 6 in (502.92 cm)
Diameter: 1 ft 2¼ in (36.19 cm)
Maximum Span: 3 ft 11 in (119.38 cm)
Function: Surface-to-Air
Weight: 1,398 lbs (634 kg)
Warhead: 119 lbs, High Explosive Fragmentation
Guidance: Radar
First Use Date: 1960
Producer: Raytheon
Users: US Army, US Marine Corps, Bahrain, Belgium, Denmark, Egypt, France, Germany, Greece, Iran, Israel, Italy, Japan, Jordan, Kuwait, Netherlands, Norway, Saudi Arabia, Singapore, South Korea, Spain, Sweden, Taiwan, United Arab Emirates, United Kingdom, Turkey, Morocco, Rumania, Republic of Korea
Other Designations: SAM-A-18
Status: Operational

The MIM-23 was a mobile launched surface-to-air missile. Named HAWK, an acronym for Homing All the Way to Kill, the MIM-23 has seen numerous upgrades and enhancements since it was first introduced in 1960. The Hawk is one of the most successful missile programs when judged by longevity and the number of operators. It has seen success on the battlefield as well: an Israeli Hawk battery is reported to have downed a MiG-25 in 1982 that was flying above the normal range of the missile. Hawks are also reported to be used as an air-launched missile by the Iranians, who when faced with a lack of replacements for the AIM-54 Phoenix used by Iranian F-14 Tomcats, modified the Hawk to fill that role.

RIM-24 Tartar

Left side view of a RIM-24 Tartar missile on launcher on the fantail of the USS *Norton Sound* (AVM-1).

View of a RIM-24 Tartar missile as it leaves the launch rail on the stern of a ship, likely the USS *Norton Sound*.

Specifications

Length: 15 ft 6 in (472.44 cm)
Diameter: 1 ft 1½ in (34.29 cm)
Maximum Span: 2 ft (60.96 cm)
Function: Surface-to-Air (shipboard)
Weight: 1,200 lb (545 kg)
Warhead: 130 lbs, High Explosive
Guidance: Radar Homing
First Use Date: 1958
Producer: General Dynamics
Users: US Navy, Australia, Germany
Status: Was operational, withdrawn from use

The RIM-24 was a single-stage short-to-medium range anti-aircraft version of the RIM-2 Terrier shipboard missile. Known as Tartar, it was the prime surface-to-air missile for smaller combat vessels, and the secondary missile after Terrier and Talos on larger ships. Ships built to carry the Tartar missile continued to be called "Tartar ships" even after the RIM-24 missile was starting to be replaced by the RIM-66 Standard Missile in the late 1960s.

"The guided missile cruiser USS *Albany* (CG-10) successfully fired three surface-to-air missiles simultaneously from forward, aft, and one side of the vessel on January 30th (1963)." The RIM-8 Talos missiles were fired from mounts at the bow and stern, while the RIM-24 Tartar missile was fired from the port side of the ship.

HGM-25 Titan I – LGM-25 Titan II

Vandenberg AFB, Calif. – Titan I (HGM-25) missile is launched. September 23, 1961.

Titan II (LGM-25) missile is launched from Cape Canaveral Air Force Station, Florida.

Titan II (LGM-25) is launched from an underground silo during a test.

Specifications (Titan II)

Length: 109 ft (3322.32 cm)
Diameter: 10 ft (304.8 cm)
Function: Surface-to-Surface ICBM
Weight: 330,000 lbs (149,823 kg)
Warhead: W38 Nuclear Warhead, 4 MT (Titan I)
Warhead: W53 Nuclear Warhead, 9 MT (Titan II)
Guidance: Inertial
First use: 1963
Producer: Martin
Users: US Air Force
Other Designations: SM-68, SM-68B
Status: Was operational, withdrawn from use

The Titan was a pair of missiles that were a mainstay of the US strategic nuclear forces. The initial variant, the Titan I was stored in a silo, but raised to the surface by an elevator for launch. This variant was designated the HGM-25. The LGM-25 Titan II was stored and launched from within its silo. The Titan II had greater range and payload lofting abilities and can be visually distinguished from the earlier version in that its upper stage is the same diameter as the lower stage. The Titan II also used hypergolic fuels, which ignited spontaneously when combined. This allowed faster reaction times for the Titan II over the Titan I. The Titan I was in service from 1962 to 1965 at Air Force bases in five states, while the Titan II was used from 1963 to 1987 at bases in three states.

The Titan II was adapted for use in the American space program, and was the launch vehicle for the Gemini spacecraft. Later versions with strap-on solid-fuel rocket motors called the Titan III, Titan IV and Titan 34 were used as space launch vehicles (SLV). As such they were not given MDS designations as their use was not a direct military function. Some Titan IIs when withdrawn from use as ICBMs were converted to use as SLVs.

AIM-26 Falcon

NUCLEAR MISSILE – The US Air Force Air Defense Command GAR-11 (later AIM-26) "Nuclear Falcon," was first air-to-air guided missile with an atomic capability. Developed by the Hughes Aircraft Company it armed ADC F-102A units. The Nuclear Falcon was being inspected by Chris M. Smith, Hughes engineering test pilot with a Convair F-102A Delta Dagger in the background of this early publicity photo.

SENTINELS OF FREEDOM – Standing at attention are members of the Falcon family of air-to-air guided missiles as seen in this publicity photo. Left to right are the radar-guided GAR-11 (AIM-26) Nuclear Falcon, the world's first air-to-air guided missile with a nuclear capability; the infrared guided GAR-2A (AIM-4C); the radar guided GAR-1D (AIM-4A); the infrared GAR-4A (AIM-4G) and the radar GAR-3A (AIM-4F). All were manufactured by the Hughes Aircraft Company, Culver City, California.

Specifications

Length: 7 ft 3 in (220.98 cm)
Diameter: 11 in (27.93 cm)
Maximum Span: 2 ft (60.96 cm)
Function: Air-to-Air
Weight: 200 lbs (90 kg)
Warhead: W54 Nuclear Warhead, 250 T
Guidance: radar
First Use Date: 1960
Producer: Hughes
Users: US Air Force
Other Designations: GAR-11, RB-27
Status: Was operational, withdrawn from use

To ensure the destruction of enemy bombers headed to targets in the United States, a variant of the Falcon missile was developed to carry a nuclear warhead. As it was determined that guidance systems then available were too inaccurate for the required hit-to-kill of a conventionally-armed warhead, a nuclear-tipped Falcon could dispatch a target with its much larger blast radius. But a nuclear warhead negated its use against low flying targets, or targets over populated areas. The AIM-26 was withdrawn from service along with the F-102 Delta Dagger that it armed. There was a conventional warhead version, the AIM-26B (used by Air National Guard F-102s), and a license built copy, the RB-27, was used by Sweden to arm the Drakken fighter.

UGM-27 Polaris

Polaris (UGM-27A) missile being launched from submarine off the coast of Cape Canaveral Air Force Station, Florida.

The first Polaris (UGM-27C) missile (test A3X-01) being prepared for launch from pad CX-29A at Cape Canaveral Air Force Station, Florida. August 7th, 1962.

Specifications (UGM-27C)

Length: 32 ft 4 in (985.52 cm)
Diameter: 4 ft 6 in (137.16)
Function: Underwater-to-Surface SLBM
Weight: 35,700 lbs (16,208 kg)
Warhead: W47 Nuclear Warhead, 1.2 MT
Guidance: inertial
First Use Date: 1959
Producer: Lockheed
Users: US Navy, United Kingdom
Other Designations:
Status: Was operational, withdrawn from use

Polaris had the distinction of being the first Submarine Launched Ballistic Missile (SLBM) of the US Navy. Using stable solid rocket fuels, the advent of the UGM-27 meant the end of missiles such as Regulus, as the missile that required less effort to fire when called upon had a certain advantage over missiles that required a submarine to surface and place the weapon into a launch position. The Polaris was fired from underwater, using a cold launch method. The missile was ejected from the launch tube within the submarine by compressed air, and as the missile cleared the surface of the water, the first stage was ignited. This system proved very successful, and was used on the missiles that succeeded the Polaris, the Poseidon and Trident. The success of the Polaris also ended in part the Navy's Martin P6M Seamaster nuclear bomber program.

Polaris (UGM-27C) missile being launched from a submarine off the coast of Cape Canaveral Air Force Station, Florida. March, 1965.

AGM-28 Hound Dog

Right side view of a Boeing B-52 Stratofortress launching an AGM-28 Hound Dog missile.

Chanute AFB, Illinois – Students being trained in the maintenance and operation of the GAM-77 (later designated the AGM-78) Hound Dog missile at the Air Training Command facility. Lt. Allen J. Partin, an instructor with the 3345th Technical School, watches as students, Airman 2nd Class William Meredith (right) and Airman 2nd Class Eugene Puck work on the Hound Dog missile trainer.

Specifications

Length: 42 ft 6 in (1295.4 cm)
Diameter: 2 ft 4 in (71.12 cm)
Height: 9 ft 4 in (284.48 cm)
Maximum Span: 12 ft 2 in (370.84 cm)
Function: Air-to-Surface
Weight: 10,147 lbs (4,607 kg)
Warhead: W28 Nuclear Warhead, 1 MT
Guidance: Inertial
First Use Date: 1959
Producer: North American
Users: US Air Force
Other Designations: WS 131B, GAM-77
Status: Was operational, withdrawn from use

The AGM-28 was an air-launched cruise missile with a range approaching 800 miles. Known as Hound Dog, it was only carried by certain versions of the B-52 Stratofortress bomber. It was termed a "stand off" weapon in that it could be launched at targets keeping the bomber well away from any local defenses. The turbojet engines of the paired AGM-28s mounted beneath the inboard wings of a B-52 could be used to supplement its takeoff power, once airborne the bomber could refuel the missiles' fuel tanks. No Hound Dogs were ever fired in anger, and they were withdrawn from use in 1978.

MGM-29 Sergeant

A MGM-29 Sergeant missile rests in level position in its trailer-launcher at White Sands Missile Range, New Mexico.

A MGM-29 Sergeant missile is fired at White Sands Missile Range, New Mexico.

Specifications

Length: 34 ft 6 in (1051.56 cm)
Diameter: 2 ft 7 in (78.74 cm)
Maximum Span: 7 ft 9 in (236.22 cm)
Function: Surface-to-Surface
Weight: 10,000 lbs (4,540 kg)
Warhead: W52 Nuclear Warhead, 200 KT
Guidance: Inertial
First Use Date: 1962
Producer: Sperry
Users: US Army, West Germany
Other Designations: SSM-A-21, XM-29
Status: Was operational, withdrawn from use

The MGM-29 was a short-range surface-to-surface missile designed for the battlefield. The Sergeant replaced the MGM-5 Corporal with Army missile troops. Developed through the 1950s the MGM-29 became operational in 1962 and deployed for the first time with Army units overseas in 1963. The MGM-52 Lance started replacing the Sergeant in 1973 and in 1977 the last battalion operating the MGM-29 was phased out of service.

LGM-30 Minuteman

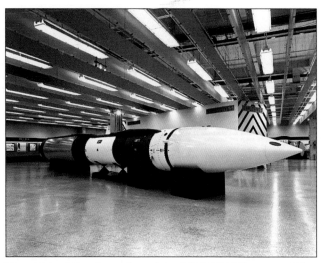

View of the full-scale mockup of the LGM-30G Minuteman III missile.

View of the LGM-30G Minuteman III missile shortly after launch.

Specifications (Minuteman III)

Length: 59 ft 10 in (1823.72 cm)
Diameter: 5 ft 6 in (167.64 cm)
Function: Surface-to-Surface ICBM
Weight: 79, 432 lbs (36,063 kg)
Warhead: W5 Nuclear Warhead, 20 KT (Minuteman I)
Warhead: W5 Nuclear Warhead, 20 KT (Minuteman II)
Warhead: (3) W5 Nuclear Warhead, 170 KT (Minuteman III)
Warhead: (1) W87 Nuclear Warhead, 330 KT (Minuteman III – current)
Guidance: Inertial
First Use Date: 1960
Producer: Boeing
Users: US Air Force
Other Designations: SM-80
Status: Operational

The LGM-30 ICBM was a step beyond previous ICBMs such as Atlas and Titan in that it used more stable solid rocket fuel in all three stages of the missile, rather than any liquid fuels such as kerosene or red fuming nitric acid. The Minuteman is hot-launched from a hardened silo; that is to say the rocket motor is ignited while the missile is still in the silo. While this protects the missile from preemptive counter fire, it prevents the immediate reuse of the silo. While the LGM-30 replaced Atlas and Titan, it too was replaced in part by the LGM-118 Peacekeeper. The Minuteman was initially fielded at six Air Force bases, but reductions in force and treaty agreements have found the last of the Minutemen missiles at now three Air Force Bases in Montana, North Dakota and Wyoming.

MGM-31 Pershing

A significant feature of the US Army's Pershing weapon system was its high mobility. Pershing (MGM-31), developed at the Martin Company's Orlando, Fla. Facility, was a two-stage, selective range, ballistic missile system. This Pershing test missile was mounted on a transporter-erector-launcher (TEL) carried by an XM474 tracked vehicle.

View of the launch of an MGM-31 Pershing missile.

Specifications (MGM-31C)

Length: 34 ft 10 in (1061.72 cm)
Diameter: 3 ft 4 in (101.6 cm)
Function: Surface-to-Surface
Weight: 16, 540 lbs (7,509 kg)
Warhead: W50 Nuclear Warhead, 400 KT (Pershing I)
Warhead: W85-W86 Nuclear Warhead, 10-50 KT (Pershing II)
Guidance: Inertial
First Use Date: 1960
Producer: Martin
Users: US Army, Germany
Other Designations: XM14
Status: Was operational, withdrawn from use

The MGM-31 missile was a battle theater ballistic missile designed to replace the PGM-11 Redstone in US Army artillery units. Named Pershing after General John J. Pershing, the Martin-built missile was capable of carrying a conventional or nuclear warhead. A two-stage missile, the MGM-31 had three main versions, the Pershing I, Ia and the Pershing II. While mainly supplied to forces in Europe, the Pershing was also supplanted in use there by numbers of the missile used by German forces. The Pershing missiles were scrapped as part of the 1988 Intermediate Range Nuclear Forces treaty. Several inert examples were retained for display and one went on display at the National Air and Space Museum alongside its Soviet counterpart, the SS-20, in 1990. Less than a year later the last Pershing unit disbanded and its missiles were crushed and destroyed.

MGM-32 ENTAC

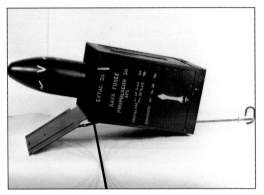

Left side view of the ENTAC (MGM-32) missile in its shipping and launch container.

View of a jeep equipped with launchers for four ENTAC (MGM-32) missiles.

Specifications

Length: 2 ft 8½ in (82.55 cm)
Diameter: 6 in (182.88 cm)
Maximum Span: 14 ft 9½ in (450.85 cm)
Function: Surface-to-Surface Anti Tank
Weight: 27 lbs (12 kg)
Warhead: 9 lbs, High Explosive, Shaped Charge
Guidance: Wire Guided
First Use Date: 1957
Producer: Nord Aviation (France)
Users: US Army, France, India, Iran, Lebanon, Australia, Belgium, Indonesia, Morocco, Netherlands, Norway, South Africa, Switzerland
Status: Was operational, withdrawn from use

The MGM-32 was a wire-guided anti-tank missile, with the operator controlling the missile in flight by means of a joystick. Known by its French acronym of ENTAC, which stood for ENgin Teleguide Anti Char, the MGM-32 had a shaped-charge warhead to pierce the armor of tanks and other vehicles on the battlefield.

The MGM-32 was normally employed by US Army troops on the M151 jeep, and was replaced by the BGM-71 TOW missile in the late 1960s.

MQM-33 Creeper Prop

Left ½ front view, from below, of an MQM-33C target drone missile on a ZL-4 launcher.

Left ½ rear view, from above, of an MQM-33 target drone missile on a ZL-4 launcher with a technician adjusting the drone.

Specifications

Length: 12 ft 7 in (383.53 cm)
Height: 2 ft 6 in (76.2 cm)
Maximum Span: 11 ft 6 in (350.52 cm)
Function: Surface Launch Aerial Target
Weight: 450 lbs (204 kg)
Guidance: Radio Control
First Use Date: 1950
Producer: Northrop
Users: US Army
Other Designations: OQ-19, BTT
Status: Was operational, withdrawn from use

A Northrop-developed Basic Training Target, or BTT, the MQM-33 was used as an aerial target for gunnery training. While the MQM-33 could be launched from a catapult, at fixed installations a circular launch track method was used. Barring a direct hit, the MQM-33 was recovered by parachute for repairs and reuse.

Dozens of MQM-33 radio-controlled aerial targets are shown readied for a day's session of anti-aircraft artillery practice at a Ft. Bliss, Texas range. Target drone on the left was on its launcher cart prior to taking off from an A-2 rotary launcher/circular strip.

BQM-34 Firebee

Left side view of the launch of an MQM-34 Firebee drone.

Right ½ front view of an MQM-34 Firebee drone mounted on the underwing pylon of a Navy Lockheed DC-130 Hercules.

Specifications (BQM-34A)

Length: 22 ft 11 in (698.5 cm)
Height: 6 ft 8½ in (204.47 cm)
Maximum Span: 12 ft 11 in (393.7 cm)
Function: Aerial Target
Weight: 2,500 lbs (1,135 kg)
Guidance: Radio Controlled
First Use or Design Date: 1958
Producer: Teledyne Ryan
Users: US Army, US Air Force, US Navy, US Marine Corps
Other Designations: Q-2
Status: Operational

Right ½ rear view of a Lockheed DC-130 Hercules in flight, with three of four underwing pylons carrying BQM-34 Firebee drones.

Perhaps one of the most successful missile programs, and certainly the most successful drone, the Firebee was used on target ranges the world over, and adapted for roles which went beyond the original variant produced. The Firebee has played roles in combat areas across the globe. Both air and ground launched, the Firebee has not only been the target of many weapons crews, versions have flown with weapons, making the Firebee the attacker. The BGM-34B could fire a variety of air-to-surface weapons, both bombs and missiles. The Firebee performed reconnaissance with both cameras and Elint (electronic intelligence) gear, and made paths across combat areas with ECM (electronic counter measure) equipment and dispensing chaff to foil radar. One variant, the Firebee II, was capable of supersonic speeds. The Firebee was certainly a workhorse capable of any task assigned.

Left ½ rear view of a US Army MQM-34D Towbee drone sitting on a maintenance dolly. The MQM-34 Firebee in this instance is carrying a crucible Infrared source on the left wingtip, so it is called a Towbee as it will be towing the target, rather than being the target itself.

View from below of a US Air Force BQM-34 Firebee drone being recovered by parachute.

Port bow view of a US Air Force missile recovery ship (MR-85-1603), operated by the 82nd Technical Assistance Team/Tactical Training Wing, returning to base with a BQM-34 Firebee drone secured to the deck. The drone was recovered from the water after completing a mission as target for aircraft participating in air-to-air combat training exercise William Tell '82.

Left ½ front view of BGM-34B Firebee drone with the assortment of weapons it could carry including the AGM-65 Maverick missile and assorted guided and high-drag bombs.

American Missiles 37

Front ¼ left view of a BQM-34S equipped with circular sensor arrays under each wing as part of the High Altitude Target-Skylite system used for laser beam testing at White Sands Missile Range, New Mexico. Laser beam data was acquired by the sensors and transmitted to ground stations during test operations.

Right ½ front view of a supersonic BQM-34E Firebee II drone ("94") mounted on the outboard pylon and a subsonic BQM-34S Firebee ("39") mounted on the inboard wing pylon of a Lockheed DC-130 transport.

"Tyndall AFB, FL – A BQM-34F Firebee II drone leaves its launch pad during air-to-air combat training exercise William Tell '82. The drone served as a target for aircraft participating in the exercise."

Left ½ front view of a BQM-34 Firebee drone just as it is released from the wing pylon of a DP-2 Neptune launch aircraft.

AQM-35

Right ½ front view of XQ-4B (AQM-35) drone mounted on the wing pylon of a Lockheed GC-130A Hercules drone carrier transport.

Left ½ front view of a Lockheed GC-130A Hercules drone carrier transport with a XQ-4B (AQM-35) drone mounted on the inner wing pylon and a Q-2C (AQM-34) mounted on the outer pylon.

Right side view of a Q-4B supersonic target drone in flight. The missile was later designated the AQM-35.

Holloman AFB, New Mexico – A Sikorsky H-34 helicopter recovering a XQ-4B test vehicle after a test flight of the missile. July 14, 1960.

Specifications

Length: 35 ft 4 in (1076.96 cm)
Diameter: 6 ft 2 in (187.96 cm)
Maximum Span: 12 ft 8 in (386.08 cm)
Function: Air Launched Aerial Target
Weight: 3,400 lbs (1,543 kg)
Guidance: Microwave Command Guidance

First Use Date: 1956
Producer: Northrop
Users: US Air Force
Other Designations: XQ-4, Q-4
Status: Tested

The AQM-35 was a high-speed aerial target developed for the US Air Force. It was capable of speeds just above one and half times the speed of sound, and was aerial launched from a DC-130 Hercules transport. Initially known as the Q-4B, the drone missile was to act as a target for surface-to-air missiles such as the MIM-14 Nike Hercules, and for air-to-air missiles like the AIM-7 Sparrow. Due in part to the speeds attained by the missile, it was recovered by a three stage parachute system. The first two parachutes in sequence would slow the craft down, and the final large parachute was used for recovery. As the AQM-35 would descend, flotation/cushion bags would deploy to soften the landing shock or keep the missile afloat in the event of a water landing.

MQM-36 Shelduck

Left ½ front view of the MQM-36 Shelduck aerial target drone on ZL-4 field launcher. The Shelduck was previously designated the KD2R-5.

View of the recovery of an MQM-36 Shelduck drone from the water. The MQM-36 is recovered by a parachute, landing in the water and a nearby ship will sail into position to hoist the drone aboard.

Specifications

Length: 12 ft 7½ in (384.81 cm)
Height: 2 ft 7 in (78.74 cm)
Maximum Span: 13 ft 2½ in (402.59 cm)
Function: Surface Launch Aerial Target
Weight: 360 lbs (163 kg)
Guidance: Radio controlled
First Use Date: 1950
Producer: Northrop
Users: US Navy
Other Designations: KD2R-1
Status: Was operational, withdrawn from use

The naval variant of the Basic Training Target, or BTT, the MQM-36 had a larger engine than the MQM-33 and had the ability to carry pods containing radar reflectors on each wingtip. The launch method for the MQM-36 was also similar to that of the MQM-33, but the Shelduck could be recovered from a water landing.

AQM-37 Jayhawk

A view of an AQM-37A target before it is uploaded onto the wing of an A-6E Intruder aircraft, at the Pacific Missile Test Center, Point Mugu flight line.

Right rear view of AQM-37 missile drone.

Specifications

Length: 12 ft 6½ in (382.27 cm)
Diameter: 1 ft 1 inch (33.02 cm)
Maximum Span: 3 ft 3½ in (100.33 cm)
Function: Target Drone
Weight: 565 lbs (256 kg)
Guidance: Programmed or Radio
First Use or Design Date: 1963
Producer: Beechcraft
Users: US Navy, US Air Force, US Army
Other Designations: Beech 1019, Beech 1102, SAGMI
Status: Was operational, withdrawn from use

An expendable high-speed aerial target, the AQM-37 was air-launched and then simulated either missiles or high-speed aircraft to train defensive systems and their operators. Although normally recovered by parachute, the Jayhawk could also self-destruct if recovery was a problem or if unsafe conditions occurred. An Air Force Armament Lab derivation of the AQM-37 was proposed as the SAGMI or Surface Attack Guided Missile, but it was not produced.

AQM-38

Left ½ front view of an RP-76 (later designated AQM-38A) drone target suspended under the wing of a Northrop F-89 Scorpion. The RP-76 was used as a high-speed target for surface-to-air missiles.

Right ½ front view of an RP-78B (later designated AQM-38B) drone target mounted on a support frame for display. Normally launched from a Northrop F-89 Scorpion. The RP-78B was used as a high-speed target for surface-to-air missiles.

Specifications

Length: 9 ft 8 in (294.64 cm)
Diameter: 1 ft (30.48 cm)
Maximum Span: 5 ft (152.4 cm)
Function: Aerial Target
Weight: 300 lbs (136 kg)
Guidance: Radio Control
First Use or Design Date: 1959
Producer: Northrop
Users: US Army, US Navy
Other Designations: RP-76, RP-78
Status: Was operational, withdrawn from use

The AQM-38 was a drone target used by the US Navy, US Air Force and the US Army to train anti-aircraft crews. Rocket powered, the AQM-38 was normally launched from a US Air Force F-89 Scorpion fighter. The Navy version had a higher impulse rocket motor, which allowed speeds for the missile of over Mach 1.

Right ½ rear view, from below, of an RP-76 (later designated AQM-38) drone target dropping from under the wing of a Northrop F-89 Scorpion. There is a second RP-76 mounted on the wing pylon of the left wing of the F-89.

MQM-39 Cardinal

Left ½ front view of a US Navy KDB-1 (later MQM-39) drone on a four-wheeled handling cart.

Right ½ front view of a US Navy KDB-1 (later MQM-39) drone mounted on the right wing pylon of a Lockheed P2V Neptune.

Specifications

Length: 15 ft 1 inch (459.74 cm)
Diameter: 1 ft 6 in (45.72 cm)
Maximum Span: 12 ft 11½ in (394.96 cm)
Function: Aerial Target
Weight: 664 lbs (301 kg)

Guidance: Radio Control
First Use Date: 1957
Producer: Beechcraft
Users: US Navy
Status: Was operational, withdrawn from use

Equipped with radar reflectors, the MQM-39 could mimic a larger aircraft to train naval anti-aircraft gunnery crews, as well as those seamen training in the use of ship-to-air and air-to-air missiles. The MQM-39, previously known as the KDB-1, could be launched from a ground launcher aided by a rocket booster, or could be pylon-mounted for launch from an aircraft.

MQM-40

Specifications

Length: 11 Feet, 6 in (350.52 cm)
Height: 1 ft 6 in (45.72 cm)
Maximum Span: 11 ft 8 in (355.6 cm)
Function: Aerial Target
Guidance: Radio Controlled

Producer: Globe
Users: US Navy
Other Designations: KD6G
Status: Was operational, withdrawn from use

The MQM-40 was an aerial target used by the US Navy for gunnery training, both surface-to-air and air-to-air. It was operated for most of its service life as the KD6G, and it was withdrawn from use in the early 1960s. The layout of the small drone was similar to many other drones, however it was equipped with a twin tail empennage, that is the fins and rudders were at the ends of the horizontal surfaces.

AQM-41 Petrel

Left side view of an AQM-41 Petrel drone mounted under the wing of a Consolidated PB4Y-2 of the Naval Aviation Ordnance Test Station, Chincoteague, Virginia.

Left ½ front view, from below, of a pair of AQM-41 Petrel drones mounted under the wings of a Lockheed P-2 Neptune in flight.

Specifications

Length: 24 ft (731.52 cm)
Diameter: 2 ft (60.96 cm)
Maximum Span: 13 ft (396.24 cm)
Function: Air-to-Surface (Sea)
Weight: 3,800 lbs (1,725 kg)
Warhead: Mark 21 Torpedo, High Explosive
Guidance: Radar
First Use Date: 1951
Producer: Fairchild Aircraft
Users: US Navy
Other Designations: AUM-N-2
Status: Was operational, withdrawn from use

Left ½ rear view of a AQM-41 Petrel drone as it is released from under the wing of a Lockheed P2V Neptune in flight.

The successful marriage of missile and torpedo technology, the AQM-41 Petrel was a Mark 21 torpedo equipped for air launch against naval targets by mating wings, tail surfaces, a guidance kit and a Fairchild J44 turbojet engine. After release from the carrier aircraft (due to size, usually a patrol bomber), the Petrel would fly a distance to the target, and at a preset altitude, the non-aquatic components would separate, allowing the torpedo to enter the water and run a course to the target vessel, either a surface ship or a submarine.

MQM-42 Redhead/Roadrunner

Right ¼ front view of an MQM-42 Redhead/Roadrunner drone undergoing checkout at Launch Complex 33, White Sands Missile Range, New Mexico.

Right side view of an MQM-42 Redhead/Roadrunner drone before launch at White Sands Missile Range, New Mexico.

Left ½ front view of an MQM-42 Redhead Roadrunner drone at launch from White Sands Missile Range, New Mexico.

Right ½ front view of an MQM-42 Redhead/Roadrunner drone after parachute recovery following launch from White Sands Missile Range, New Mexico.

Specifications

Length: 19 ft (579.12 cm)
Diameter: 1 ft (30.48 cm)
Maximum Span: 6 ft (182.88 cm)
Function: Ground Launched High Speed Target
Guidance: Radio Controlled
First Use Date: 1960
Producer: North American Aviation
Users: US Army
Other Designations: NA273
Status: Was operational, withdrawn from use

The MQM-42 Redhead/Roadrunner was a bit unusual in having two jointed names, but this perhaps was to indicate that this aerial target was intended to operate at both high and low altitudes, and at subsonic and supersonic speeds. The MQM-42 was built as a target for the MIM-23 Hawk and MIM-14 Nike Hercules air-defense missiles.

FIM-43 Redeye

Private James Meching sighting the FIM-43 Redeye missile on a ballistic aerial target.

Troops at Fort Riley, Kansas fire the FIM-43 Redeye missile at a moving target.

Specifications

Length: 48 in (121.91 cm)
Diameter: 3 in (7.61 cm)
Function: Surface-to-Air Missile
Weight: 29 lbs (13 kg)
Warhead: 2.5 lbs, High Explosive, Fragmentation
Guidance: Infrared
First Use Date: 1961
Producer: General Dynamics
Users: US Army, US Marine Corps, Sweden, Australia, Croatia, El Salvador, Germany, Israel, Nicaragua, Thailand, Turkey
Status: Was operational, withdrawn from use

Having determined that the target was within range, the Army gunner pulled the trigger that sent the FIM-43 Redeye air-defense guided missile on its way. The Redeye is shown here just after it cleared its shoulder launcher. When the missile was a safe distance from the gunner, its sustainer motor fired and powered it to the target.

The Redeye was a shoulder-fired surface-to-air missile of the type termed "fire and forget." The soldier firing the FIM-43 would sight his target, the missile would signal that it had a lock with the infrared seeker head, the missile would be launched and while it tracked to the target at supersonic speed, the soldier could move to cover or pick up another weapon to fire against a second aerial target. The Redeye had a two stage motor, an ejection motor to force the missile from the launch tube and away from the operator, and a stronger sustainer motor that would send the missile at high speed to the target.

UUM-44 Subroc

Left ½ front view of a UUM-44 Subroc missile on a dolly. Unidentified Goodyear employee stands next to the missile.

Right side view of a submarine launched UUM-44 Subroc anti-submarine missile as it leaves the water during launch.

Specifications

Length: 21 ft (640.08 cm)
Diameter: 21 in (53.33 cm)
Function: Underwater-to-Underwater Anti-submarine Missile
Weight: 4,000 lbs (1,816 kg)
Warhead: W55 Nuclear Warhead, 1-5 KT
Guidance: Inertial
First Use Date: 1959
Producer: Goodyear Aerospace
Users: US Navy
Status: Tested

Right front side view of a submarine launched UUM-44 Subroc anti-submarine missile as it leaves the water during launch.

In a game of Cold War cat-and-mouse, fast attack submarines pursued ballistic missile submarines, which were armed with missiles to attack cities, military installations and other targets of the opposing side. A weapon carried by fast attack submarines of the US Navy, the UUM-44 Subroc was a weapon intended for use against those other submarines. With longer range than a conventional torpedo, the Subroc was initially similar in that it was to be fired horizontally from a conventional torpedo tube. After leaving the submerged vessel, the UUM-44 would arc upward, leave the water and accelerate to supersonic speeds. As the main rocket motor burned out, separation motors would ensure that the booster and warhead would part company. The booster would tumble to the ocean, while the warhead/nuclear depth charge would continue on a guided trajectory to the target submerged miles away from the launch submarine. The warhead would enter the water, descend to the programmed depth and explode.

AGM-45 Shrike

Left side view of an AGM-45A Shrike missile on a support stand. Behind the AGM-45A is an AIM-9E Sidewinder on a weapons trailer.

Right ½ front view of an AGM-45 Shrike missile.

Specifications

Length: 10 ft (304.8 cm)
Diameter: 8 in (20.32 cm)
Maximum Span: 3 ft (91.44 cm)
Function: Air-to-Surface
Weight: 390 lbs (177 kg)
Warhead: 145 lbs, High Explosive Fragmentation
Guidance: Passive Radar Seeker
First Use Date: 1963
Producer: Texas Instruments
Users: US Navy, US Air Force
Status: Was operational, withdrawn from use

The AGM-45 was an air-to-surface anti-radar missile used to combat enemy air defense systems. The Shrike homed in on the signal transmitted by ground tracking radars, with the intent of destroying enemy radar before anti-aircraft weapons could be brought to bear. The Shrike was one of the main weapons used by US Air Force "Wild Weasel" F-105G and F-4G aircraft on radar-suppression missions, and was also used on US Navy carrier based Iron Hand aircraft including the A-4, A-6, A-7 and F/A-18. The AGM-45 was developed by the Naval Ordnance Test Station, China Lake, California, and was retired from use after Operation Desert Storm.

MIM-46 Mauler

Artist's concept of a tracked vehicle equipped with a turret for firing the MIM-46 Mauler missile.

Rear view of a XM546 tracked vehicle equipped with a turret for firing the MIM-46 Mauler missile as soldiers load the launcher with missiles.

Specifications

Length: 6 ft (182.88 cm)
Diameter: 5 in (12.7 cm)
Maximum Span: 1 ft 6 in (45.72 cm)
Function: Surface-to-Air
Weight: 119 lbs (54 kg)
Warhead: High Explosive
Guidance: Radar, infrared in terminal phase
Design Date: 1961
Producer: General Dynamics
Users: US Army
Status: Unbuilt

Left 1/2 front view of an XM546 tracked vehicle equipped with a turret for firing the MIM-46 Mauler missile as soldiers enter the vehicle.

A surface-to-air missile, the MIM-46 was to provide a mobile air defense system against high-speed low-flying aircraft, and battlefield missiles for Army units in the field. The Mauler was designed to use the M546 carrier, which was based on the same M113 chassis that was in use with Army units. Problems with the development of a weapons system with a vehicle mounting both the tracking and guidance radar with the weapon itself led to project cancellation in July 1965. Lack of the MIM-46 was taken up by improvements in the MIM-23 Hawk and development of the MIM-72 Chaparral.

AIM-47 Falcon

MISSILES OF TOMORROW – The Air Force AIM-47A, at left, is designed for use with the YF-12A interceptor; on the right the AIM-54A Phoenix, for the Navy's F-111B. At center is Hughes test pilot Chris M. Smith.

Front ¼ left view of a AIM-47 Falcon missile parked on a dolly by the nose of a Lockheed YF-12A Blackbird interceptor.

Specifications

Length: 12 ft 6 in (381 cm)
Diameter: 13½ in (34.29 cm)
Maximum Span: 33 in (83.82 cm)
Function: Air-to-Air Missile
Weight: 800 lbs+ (363 kg)
Warhead: W42 Nuclear Warhead
Guidance: Radar
First use: 1962
Producer: Hughes
Users: US Air Force
Other Designations: GAR-9
Status: Tested

Another variant of the Falcon family, the AIM-47 was destined to never see service, as the aircraft it was designed to arm was never placed into production. Intended as the air-to-air weapon of the stillborn North American F-108 and then the Lockheed YF-12A "Blackbird" interceptor, the nuclear-tipped AIM-47 would have been targeted at the expected Soviet bomber forces that would have approached the North American continent. Among the special needs for this version of the Falcon were the ability to be withstand a stable ejection from the three missile bays of an aircraft traveling in excess of Mach 2 and continue on to track and destroy a target while flying at speeds in excess of Mach 4. Some of the design elements that went into engineering the AIM-47 were put to use in the long-range AIM-54 Phoenix missile used on the Grumman F-14 Tomcat.

AGM-48 Skybolt

Right side view of an AGM-48 Skybolt missile on a handling dolly, parked in front of the nose of a Boeing B-52 Stratofortress.

Left side view, from below, of four AGM-48 Skybolt missiles suspended under pylons on the wings of a Boeing B-52 Stratofortress.

Specifications

Length: 38 ft 3 in (1165.86 cm)
Diameter: 2 ft 11 in (88.9 cm)
Maximum Span: 5 ft 6 in (167.64 cm)
Function: Air-to-Surface
Weight: 11,000 lbs (4,994 kg)
Warhead: W59 Nuclear Warhead, 1.2 MT
Guidance: Inertial
First Use Date: 1962
Producer: Douglas
Users: US Air Force
Other Designations: WS 138A, GAM-87
Status: Tested

The Skybolt was an abortive attempt at using an aerial platform to disperse a fleet of ALBM, or Air Launched Ballistic Missiles, with the aim that such a method would make them less vulnerable to preemptive attack. Using the launch bomber's navigation system to update the internal inertial navigation system on the missile made the concept feasible, and the B-52 was planned to carry a load of four of the AGM-48s. The RAF was also to use the Skybolt: a Vulcan bomber could carry one of the missiles aloft. In joining the development of the AGM-48, the British Blue Streak and Blue Steel missiles were cancelled. Skybolt was a two-stage vehicle, with an aerodynamic shroud on the rear while under transport. At launch the tailcone would be jettisoned as the motor ignited, and the first stage fins would steer the missile. At first stage burnout, a gimbaled nozzle would steer the remaining missile and warhead to the target.

But the test program to prove Skybolt was without success in 1962 until December, on the same day the plug was pulled on the program. The cancellation led to a stronger reliance on the submarine-launched Polaris, and while the US Air Force could still rely on improved silo-based missiles, the RAF was left holding an empty bag as the strategic nuclear responsibilities for Great Britain were turned over to the Royal Navy and their nuclear submarine fleet.

LIM-49 Nike-Zeus

Right ½ front view of a LIM-49 Nike-Zeus A missile on launch rail. "White Sands Missile Range, N.M. – The US Army Nike-Zeus missile erected on launching rail in preparation for its first test firing at Army Launching Area 5."

Left ½ front view of a LIM-49 Nike-Zeus B anti-missile missile on launch rail at Point Mugu Missile Range, California. Dated September 8, 1961.

Specifications (Nike-Zeus B)

Length: 32 ft 6 in (990.6 cm)
Diameter: 3 ft 9 in (114.3 cm)
Maximum Span: 9 ft 9 in (297.18 cm)
Function: Surface-to-Air
Weight: 22,000 lbs (9.988 kg)
Warhead: W71 Nuclear Warhead, 5 KT
Guidance: Radar Guided
First Use Date: 1959
Producer: Western Electric/Douglas
Users: US Army
Other Designations:
Status: Tested

Left ½ front view of a LIM-49 Nike-Zeus B missile leaving launch rail. "White Sands Missile Range, N.M. – Launch of the Nike-Zeus missile #20018 at Army Launching Area 5."

The LIM-49 missile was developed to intercept re-entry vehicles from enemy ballistic missiles, but its role changed after the launch of Sputnik in 1957. The initial variant, the Nike-Zeus A was a two-stage missile that looked somewhat like an upscaled Nike-Hercules. But at that time the Army was restricted in the range of its defensive missiles. Shortly after the launch of Sputnik in October of 1957, the LIM-49 was redesigned as the Nike-Zeus B. The powerful first stage was retained, but two new upper stages were added, revamping the look of the missile. With the range increase, the Nike-Zeus B was not just targeting warheads and missiles as they reentered the atmosphere; it was also targeting objects in orbit above the atmosphere, including satellites. The Nike-Zeus program was canceled before it went into full production and employment; however, the development of Nike-Zeus played a role in the later Spartan missile program.

RIM-50 Typhon LR

Long Range Typhon serial No. 1 prior to being fired by the US Naval Ordnance Test Facility at White Sands Missile Range, N.M. Left ½ front view of RIM-50 Typhon LR missile on launcher.

Long Range Typhon serial No. 1 is fired by the US Naval Ordnance Test Facility at White Sands Missile Range, N.M. Right ½ rear view of RIM-50 Typhon LR missile as it leaves the launcher.

Specifications

Length: 46 ft (1402.08 cm)
Function: Surface-to-Air
Weight: 20,000 lbs (9,080 kg)
Warhead: High Explosive
Guidance: Radar
First Use Date: 1961
Producer: Bendix
Users: US Navy
Other Designations: SAM-N-8, Advanced Talos
Status: Tested

Developed by the Applied Physics Laboratory of Johns Hopkins University, the Typhon was a further derivation of the "Bumblebee" program as were the earlier Terrier, Tartar and Talos; but Typhon was to replace the other three. At one point, the RIM-50 was known as Advanced Talos. Typhon LR was to be a long-range surface-to-air missile as a defensive weapon for surface ships. The MIM-50 program was cancelled in late 1963.

MGM-51 Shillelagh

ARMY'S NEW TANK KILLER – Shillelagh (MGM-51), the Army's first guided missile to be fired from a tank is launched from the M551 Sheridan Armored Reconnaissance Airborne Assault Vehicle. Guided by an automatic command system mounted on the launching vehicle, the missile flies along the tank gunner's line of sight to the target. Highly accurate against stationary or moving target, Sheridan's 152mm gun launcher can fire either MGM-51 missiles or conventional ammunition. October 1967.

US Army soldier handles modern Shillelagh (MGM-51), anti-armor and field fortification missile that packs a far greater wallop than the one carried by its fabled Irish namesake. Under development by the Aeronutronic Division of Philco Corporation, a subsidiary of Ford Motor Company, Newport Beach, California for Sheridan/Shillelagh Project manager, Army Weapons Command, Rock Island, Illinois. May 18, 1964.

Specifications

Length: 3 ft 7 in (109.22 cm)
Diameter: 6 in (15.23 cm)
Maximum Span: 12 in (30.47 cm)
Function: Surface-to-Surface Anti-tank
Weight: 40 lbs (18 kg)
Warhead: 13 lbs, High Explosive, Shaped Charge
Guidance: Command
First Use Date: 1963
Producer: Philco Ford, Martin Marietta
Users: US Army
Other Designations: XM-13
Status: Was operational, withdrawn from use

Head-on view as a MGM-51 Shillelagh missile leaves the barrel of an M551 Sheridan tank. "Close-up of the Sheridan Shillelagh Weapons System being fired at Gun Position 17 at the Yuma Proving Ground, during test firing phases."

Designed as the 152mm missile round to be fired from the M551 Sheridan tank's gun launcher, the Shillelagh was also projected to be the weapon of choice for the proposed Main Battle Tank, the MBT-70. The MBT-70 was not produced, although the MGM-51 was adapted for use on some M-60 tanks. An effort was made for the Shillelagh to be adapted to perform as an infantry weapon instead of the BGM-71 TOW, but the Shillelagh did not meet the requirements that the TOW was meeting for an infantry anti-tank weapon. Some documents indicate a projected nuclear capable Shillelagh, but the range (1¼ miles for the MGM-51A, 1⅞ miles for the MGM-31B and C) made this unlikely.

MGM-52 Lance

White Sands Missile Range, N.M. – A view showing the MGM-52 Lance missile being elevated in preparation for firing during a tactical test. November 3, 1965. The MGM-52 Lance is being carried in a M752 TER (Transporter Erector Launcher).

Right ½ front view showing the MGM-52 Lance missile being prepared for firing during tactical training. Boeing CH-47 Chinook helicopter in the background.

Specifications

Length: 20 ft (609.6 cm)
Diameter: 1 ft 10 in (55.87 cm)
Function: Surface-to-Surface
Weight: 3,200 lbs (1,453 kg)
Warhead: 1,000 lbs High Explosive, W70 Nuclear (up to 100 kT)
Guidance: Inertial
First Use Date: 1965
Producer: Vought
Users: US Army, Israel, United Kingdom, Netherlands, Belgium, Italy, Germany
Other Designations: Missile B
Status: Was operational, withdrawn from use

Right ½ front view showing two MGM-52 Lance missiles being fired during a tactical training exercise.

The MGM-52 was capable of delivering a nuclear or conventional warhead. The Lance replaced the Army's Honest John, Little John and Lacrosse weapons in supplanting conventional artillery with the added range that a missile could provide. In addition to the mobility offered for the Lance when mounted on a M752 TER (Transporter Erector Launcher), the MGM-52 was air-transportable by some of the larger Army helicopters.

Following the stand-down of the MGM-52 from front-line combat units in 1992, the Lance found use as targets for anti missile systems.

AGM-53 Condor

Right ¼ front view of an AGM-53A Condor missile mounted on the left pylon of a Grumman A-6 Intruder attack aircraft at the Naval Weapons Center, China Lake, California.

Left ½ front view, from below, of a AGM-53 Condor missile.

Specifications

Length: 13 ft 10 in (421.64 cm)
Diameter: 1 ft 5 in (43.16 cm)
Maximum Span: 4 ft 5 in (134.62 cm)
Function: Air-to-Surface
Weight: 2,100 lbs (953 kg)
Warhead: 630 lbs High Explosive shaped charge
Guidance: Television
First Use or Design Date: 1970
Producer: Rockwell
Users: US Navy
Status: Tested – cancelled

Right side view of a AGM-53A Condor missile with nose just about to touch a target bridge at the Naval Weapons Center, China Lake, California.

The Condor was a television-guided air-to-surface stand-off missile designed for use by naval attack aircraft such as the A-6 Intruder and F-111B. Both of these aircraft carried Bombardier-Navigators who would use the cockpit television display to view the image transmitted by the missile to guide the AGM-53 to its target. In the single-seat Corsair II, the pilot would be "multi-tasking" as he flew the aircraft and guided the missile. It was fired for the first time by an F-4 Phantom II. The AGM-53 program was cancelled in 1976 because of extreme cost overruns that would have made it cost four times as much as the AIM-54 Phoenix.

AIM-54 Phoenix

Hughes Aircraft Company production workers move an AIM-54C+ Phoenix missile into position for testing at Hughes' manufacturing facility in Tucson, AZ. Additional Phoenix air-to-air missiles are lined up awaiting tests in the plant's final assembly and check-out (FACO) area, where propulsion units and ordnance are installed. After completing tests at FACO, the AIM-54 missiles are delivered to the US Navy to provide long-range fleet air defense aboard F-14 Tomcat fighter aircraft.

Left side view, from below, of a Grumman F-14 Tomcat fighter armed with six AIM-54 Phoenix missiles as it climbs vertically. (Note cloud horizon below engine exhausts.)

Specifications

Length: 13 ft (396.24 cm)
Diameter: 15 in (38.1 cm)
Maximum Span: 3 ft (91.44 cm)
Function: Air-to-Air
Weight: 1,024 lbs (465 kg)
Warhead: 135 lbs high explosive fragmentation
Guidance: Radar
First Use Date: 1966
Producer: Hughes
Users: US Navy, Iran
Status: Was operational, withdrawn from use

The AIM-54 Phoenix missile was first developed for use on the F-111B, the swing-wing fleet defense fighter that the Navy was intending to field. When the Navy dropped out of the F-111 program, and purchased the Grumman F-14 Tomcat as the primary fleet defense fighter, the AIM-54 became the primary weapon for that aircraft. First flown unguided on April 27, 1966 and in a guided mode on May 12, 1966, these initial flights were made from an A-3A Skywarrior in preparation for flight tests from the F-111B which initially took place in March of 1967. Following withdrawal from the F-111 program and production of the Tomcat, the F-14 live-fired a Phoenix for the first time in August of 1970. The F-14 carried up to six AIM-54 missiles at a time. The AWG-9 radar housed in the nose of the F-14 was a key part of the weapons system, with an ability to track up to 24 targets, six simultaneously. From what open sources state, no American F-14 has fired an AIM-54 in anger. In January 1974 when Iran signed a contract for the F-14A with the United States, it included an undisclosed number of AIM-54A Phoenix air-to-air missiles as part of the purchase. Iran was reported at the time to be buying the F-14A as a counter to Soviet overflights of that country by MiG-25 reconnaissance aircraft. The MiG-25 flew at altitudes well beyond Iranian defensive range, but the Tomcat armed with the Phoenix would negate that advantage. The Phoenix was developed from extensive prior programs from Hughes, including various versions of the Falcon missile and other programs that did not reach production status.

After the Iranian Revolution, the AIM-54A was compromised by Iranian source diverting examples to the USSR, leading to a similar missile arming the MiG-31 (which the Soviet pilots refer to as the Phoenix!). This led to the introduction of the improved AIM-54C. The AIM-54C+, used by the F-14D, was similar to the -54C but did not require the older missile's liquid cooling system.

RIM-55 Typhon MR

Specifications: None

Project cancelled: No known images of this missile.
Other Designations: SAM-N-9

Derived from RIM-24 Tartar.

The RIM-55 was related to the RIM-50 and both were cancelled in late 1963 for extreme costs. Both were capable missiles, but it was felt that too few "Typhon" ships could be built, so that taking those out would expose the naval battle group to air attack.

PQM-56

Left 1/2 front view of a CT.41 drone on a work platform.

Left side view of a CT.41 drone as it is launched.

Specifications

Length: 32 ft (975.36 cm)
Diameter: 1 ft 8 in (50.8 cm)
Maximum Span: 12 ft (365.76 cm)
Function: Aerial Target Drone
Weight: 2,865 lbs (1,300 kg)
Guidance: Radio Control
First Use Date: 1959
Producer: Bell
Users: US Navy
Other Designations: Nord CT.41
Status: Tested

The Nord CT.41 was a supersonic ramjet target drone which could emulate a high-speed bomber aircraft. Originally produced in France by Nord as the CT.41, it was built by Bell Aircraft in the United States as the PQM-56 for the US Navy.

Left 1/2 rear view of a CT.41 drone elevated to launch position. The CT.41 was built by Bell Aircraft as the PQM-56.

MQM-57 Falconer

Right side view of an SD-1 (later designated the MQM-57) surveillance drone on launching stand. Note the underwing rocket motors to assist take off.

Left side view of an SD-1 (later designated the MQM-57) surveillance drone taking off.

Specifications

Length: 13 ft 5 in (408.94 cm)
Height: 2 ft 7 １/３ in (79.57 cm)
Maximum Span: 11 ft 6 in (350.52 cm)
Function: Surveillance Drone
Weight: 466 lbs (211 kg)
Guidance: Radio Control
First Use Date: 1955
Producer: Northrop
Users: US Army
Other Designations: RP-71, AN/USD-1, SD-1
Status: Was operational, withdrawn from use

Derived from the BTT series which included the MQM-33, the MQM-57 was a slightly larger vehicle as it was used for battlefield reconnaissance, not as a target drone. Launched from a portable stand, initiation of flight was aided by underwing rocket motors which dropped away after the drone became airborne. While tracked by radar, control was by radio so that the operator could acquire a subject and take either photographs or use an onboard television camera.

MQM-58 Overseer

Left side view of a SD-2 (later designated the MQM-58) surveillance drone on its launcher trailer. The cage wrapped around the spinner on the nose of the drone is the starting mechanism.

Improved version of the SD-2 (later designated the MQM-58) surveillance drone takes off.

Specifications

Length: 16 ft 1 inch (490 cm)
Height: 3 ft 7 in (109 cm)
Maximum Span: 13 ft 4 in (406 cm)
Function: Surveillance Drone
Weight: 992 lbs (450 kg)
Guidance: Radio Control
First Use Date: 1957
Producer: Rheem, Aerojet General
Users: US Army
Other Designations: AN/USD-2, SD-2, XAE-2
Status: Was operational, withdrawn from use

A reconnaissance drone much like the MQM-57, the MQM-58 Overseer was also about the same size. A visual difference with the MQM-57 was that the MQM-58 was built with a "vee" tail, rather than the traditional horizontal and vertical surfaces. The Overseer was launched for flight over the battlefield from a portable stand, which could also be mounted on a vehicle or trailer. Solid-fuel rocket motors aided the drone in getting airborne, and shortly after burnout were dropped.

RGM-59

Specifications: None

Project cancelled: No known images of this missile.

A fleet air defense missile, the RGM-59 was based on the Terrier, but with a 1,000lb warhead.

AQM-60 Kingfisher

The XQ-5 (later designated the AQM-60) is readied for flight by Lockheed Missiles and Space Co, Inc. technicians. The XQ-5 was a modified X-7A3 which substituted a pair of rocket boosters, one under each wing, for the tandem boosters used in the original X-7. The XQ-5 was a target drone to test newly-developed interceptor missiles.

Left ½ front view of an XQ-5 (later designated the AQM-60) on a ground handling dolly, with a technician alongside.

Specifications

Length: 38 ft (1158.24 cm)
Height: 7 ft (213.36 cm)
Maximum Span: 10 ft (304.8 cm)
Function: high speed target drone
Weight: 8,900 lbs (4,040 kg)
Guidance: Programmed w Radio Control
First use date (X-7): 1951
Producer: Lockheed
Users: US Air Force
Other Designations: X-7, XQ-5
Status: Tested

With a top speed of over Mach 4, the AQM-60 was indeed a high-speed target drone. Intended as a target for high-speed missiles such as the MIM-10 Bomarc and MIM-14 Nike-Hercules, the problem with the Kingfisher was that it was faster than the missiles attempting to shoot it down. While the X-7 had a long career as a ramjet test vehicle, the follow-on as a target drone did not, mainly due to the incompatibility of the drone with the targeting missiles. The guidance of the drone was by a preset program, which had an override available for control inputs and deployment of the recovery system. As in the X-7, recovery was by descent under a parachute until the nose spike/antenna entered the ground.

MQM-61 Cardinal

Right ½ front view of a US Army KDB-1 (later MQM-61) drone on a launch stand. Note the rocket motors mounted to the bottom of the rear fuselage to assist take off.

Right ¼ front view of a US Army KDB-1 (later MQM-61) drone in flight.

Specifications

Length: 15 ft 1 inch (459.74 cm)
Diameter: 1 ft 6 in (45.72 cm)
Maximum Span: 12 ft 11½ in (394.96 cm)
Function: Aerial Target
Weight: 664 lbs (301 kg)
Guidance: Radio Control
First Use Date: 1957
Producer: Beechcraft
Users: US Army
Status: Was operational, withdrawn from use

The MQM-61 was a target drone that was used to simulate larger aircraft to train soldiers who manned the Hawk anti-aircraft missile batteries. The Cardinal, previously known as the KDB-1, was launched from a ground position aided by a rocket booster. The MQM-61 was also used to tow targets and to carry "near-miss" recording devices. With the addition of heat sources, the Cardinal could also train defending infrared-seeking missiles like the Redeye and Chaparral. The MQM-61 was used up until 1977.

AGM-62 Walleye

Left side view of a Douglas A-4 Skyhawk in flight carrying an AGM-62 Walleye missile on the left pylon.

Right 1/2 front view of an AGM-62 Walleye missile on a dolly, ready for loading for testing at the Naval Weapons Center, China Lake, California. In the background is a McDonnell F-4B Phantom II.

Specifications (Walleye I)

Length: 11 ft 4 in (345.44 cm)
Diameter: 1 ft 3 in (38.1 cm)
Maximum Span: 3 ft 9 in (114.3 cm)
Function: Air-to-Surface
Weight: 1109 lbs (503 kg)
Warhead: 825 lbs, High Explosive (Walleye I)
Warhead: 2,000 lbs, High Explosive (Walleye II)
Guidance: Television
First Use Date: 1963
Producer: Martin, Hughes
Users: US Navy. US Air Force
Other Designations: Guided Weapon Mark 1 Mod 0
Status: Was operational, withdrawn from use

The Walleye was another weapon that was given a missile designation even though it was unpowered, making it more guided bomb than missile. But the AGM-62 was developed and tested by the US Navy as a television-guided weapon, with decent stand off range even as a glide bomb. A slightly smaller version, named SNIPE, for use on Army helicopters was in development behind the Walleye. SNIPE was cancelled, but the Walleye was employed in Vietnam, although the 1,100-lb shaped charge warhead had minimal effect on some hardened targets, such as the Thanh Hoa Bridge in North Vietnam. Walleye was developed at the Naval Weapons Center at China Lake, California, and pilot production of a small number of the weapons was done at the Naval Avionics Facility, Indianapolis before full-scale production began. The Walleye was a family of air-to-surface glide bombs. The 1,000-lb class variants included the Walleye I (Mk 1 with a conventional warhead, and USAF Mk 6 with a nuclear warhead) and Walleye I Extended Range/Data Link (ERDL). These were the Mk 21. Mk 29 and Mk 34. The 2,000-lb variants included the Walleye II (Mk 5) and the Walleye II ERDL (Mk 23, Mk 30 and Mk 37) Both the Mk 1 and Mk 5 were used in Vietnam and few Walleyes were expended on targets during Operation Desert Storm. The AGM-62s were withdrawn from service in the mid-1990s.

AGM-63 Brazo / Pave Arm

Specifications

Similar to AIM-7 Sparrow
Program cancelled

Brazo was a Hughes/US Navy project begun in 1972 to develop a missile that would passively home on the radar emissions of an enemy aircraft (the MiG-25 Foxbat). In 1973 this effort was joined with the USAF PAVE ARM program, which may have wanted to use the missile against both aircraft and ground-based radars. Based on the AIM-7 Sparrow, but fitted with a broad-band passive homing seeker, flight trials using F-4 "shooters" and BQM-34 targets began in 1974 at the Pacific Missile Test Center. The program was cancelled in 1978 in favor of the AGM-88 HARM.

AGM-64 Hornet

Specifications

Length: 6 ft 2 in (187.96 cm)
Diameter: 11 in (27.94 cm)
Maximum Span: 1 ft 9 in (53.34 cm)
Function: Air-to-Surface
Weight: 225 lbs (102 kg)
Warhead: High Explosive
Guidance: Electro-Optical (Television)
First use: 1964
Producer: Rockwell
Users: US Air Force
Other Designations: ATGAR
Status: Tested

Right side view of an AGM-64 Hornet missile on a pylon under the wing of a QF-100F Super Sabre fighter.

Although the AGM-64 Hornet tested several guidance methods for accurate targeting, the weapons system is most noted for the use of electro-optical guidance or television guidance. The air-launched anti-tank Hornet never achieved production, but was built in sufficient numbers to allow testing not only for the AGM-64, but also to test methodologies for other missile systems.

AGM-65 Maverick

Left ½ front view of four AGM-65 Maverick missiles laid out for display with a technician. From front to back they are the AGM-65D Infrared-guided Maverick, the laser-guided AGM-65E, the television-guided AGM-65B and lastly the AGM-65F also infrared-guided, but modified for anti-ship use by the US Navy.

F-16 MAVERICK LAUNCH – A television-guided AGM-65B Maverick missile is fired from a General Dynamics F-16 Fighting Falcon fighter over China Lake, California. During the flight, the aircraft was piloted by US Air Force Lt. Col. Tuck McAtee with Maj. Sven Hjort of the Royal Danish Air Force in the rear seat. During this flight, the aircraft was loaded with six Mavericks.

Specifications

Length: 8 ft 1 inch (246.38 cm)
Diameter: 1 ft (30.48 cm)
Maximum Span: 2 ft 4 in (71.12 cm)
Function: Air-to-Surface Guided Missile
Weight: 460 to 804 lbs (209-365 kg)
Warhead: 125 lbs, High Explosive Shaped Charge or 300 lbs, High Explosive
Guidance: Infrared, Laser and Optical
Acceptance Date: August 1972
Producer: Hughes, Raytheon
Users: US Air Force, US Navy, US Marine Corps, Belgium, Canada, Chile, Czech Republic, Denmark, Greece, Hungary, Indonesia, Iran, Italy, Japan, Jordan, Malaysia, Morocco, Netherlands, New Zealand, Pakistan, Poland, Portugal, Serbia, Singapore, Republic of Korea, Spain, Sweden, Taiwan, Thailand, Turkey, United Kingdom
Status: Operational

The Maverick is an air-to-surface guided missile that uses a modular capability to adapt the weapon to the intended target. There are three types of guidance noses for the AGM-65, laser, optical and infrared. There are two warheads for the Maverick, a 125-lb shaped charge for armored targets and a 300-lb blast warhead for use against large hard targets such as fortifications. The solid rocket motor that propels the AGM-65 upon release from the aircraft is common to all versions of the missile. Hughes proposed another variant of the Maverick using millimeter wave radar to seek air defense radars, such as used at anti-aircraft missile batteries. During the War on Terror, the AGM-65E proved to be the most useful as it could be precisely guided to the target that was laser-illuminated by troops on the ground. Built in the least numbers, a contract was issued to convert other Mavericks to the AGM-65E standard in 2009.

RIM-66 Standard Missile

Right side view of an RIM-66 Standard MR (Medium Range) missile on a Mark 26 launcher, prior to being fired from the Aegis guided missile cruiser USS *Ticonderoga* (CG-47) during tests near the Atlantic Fleet Weapons Training Facility, Roosevelt Roads, Puerto Rico. Dated March 12, 1983.

View as an RIM-66 Standard MR (Medium Range) missile is fired from a Mark 26 launcher aboard the Aegis guided missile cruiser USS *Ticonderoga* (CG-47) during tests near the Atlantic Fleet Weapons Training Facility, Roosevelt Roads, Puerto Rico. Dated May 1, 1983.

Specifications

Length: 14 ft 8 in (447.04 cm)
Diameter: 1 ft 1½ in (34.29 cm)
Maximum Span: 3 ft 5 in (104.14 cm)
Function: Surface-to-Air
Weight: 1,370 lbs (622 kg)
Warhead: High Explosive Fragmentation
Guidance: Radar
First Use Date: 1965
Producer: General Dynamics/Ratheon
Users: US Navy
Other Designations: SM-1 MR, SM-2 MR
Status: Operational

The RIM-66 was a missile designed to replace the RIM-24 Tartar missile. As the medium-range variant of the Standard Missile, it was flown with a single stage from its shipboard launchers. The extended-range Standard Missile was flown with a booster, and this two-stage version is the RIM-67.

RIM-67 Standard ER Missile

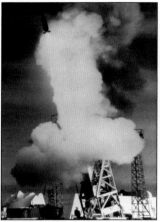

View of RIM-67 Standard ER (Extended Range) missiles being loaded on the missile house rail aboard the guided missile destroyer USS *Mahan* (DDG-42). The missiles were fired as part of the test and evaluation of the New Threat Upgrade (NTU) Combat System.

White Sands Missile Range, New Mexico – An RIM-67 Standard ER (Extended Range) surface-to-air missile being launched from the Vertical Launching System (VLS) during testing.

White Sands Missile Range, New Mexico – An RIM-67 Standard ER (Extended Range) surface-to-air missile (left), just launched from the Vertical Launching System (VLS), approaches a BQM-34 Firebee target drone during testing.

Specifications

Length: 26 ft 2 in (797.56 cm)
Diameter: 1 ft 1½ in (34.29 cm)
Maximum Span: 3 ft (91.44 cm)
Function: Surface-to-Air
Weight: 1,558 lbs (707 kg)
Warhead: High Explosive Fragmentation

Guidance: Radar and Inertial
First Use or Design Date: 1978
Producer: Raytheon
Users: US Navy
Other Designations: SM-1 ER, SM-2 ER
Status: Was operational, withdrawn from use

The Standard Missile SM-2 ER (Extended Range) is a replacement for the RIM-2 Terrier missile in the US Navy. Using the normal Standard Missile upper stage with a long first stage, the RIM-67 does not fit on Aegis Missile ships, and because the missile is ship specific to those that carried the RIM-2, they are often still called "Terrier Ships."

AIM-68

Specifications

Function: Air-to-Air
Warhead: High Explosive
Guidance: Infrared
Design Date: 1965

Producer: Air Force Weapons Lab/General Dynamics
Users: US Air Force

The AIM-68 was developed as a weapon to replace the unguided nuclear-tipped AIR-2 Genie rocket. The AIM-68 program was cancelled in 1966.

AGM-69 SRAM

View from left below of a General Dynamics F-111 with a load of four AGM-69 SRAM missiles mounted on the weapons pylons.

Right ½ front view, from above, of a workshop floor with five AGM-69 SRAM missiles undergoing checks.

Specifications

Length: 14 ft (426.72 cm)
Diameter: 1 ft 5 in (43.18 cm)
Maximum Span: 2 ft 6 in (76.2 cm)
Function: Air-to-Surface Missile
Weight: 2,208 lbs (1,002 kg)
Warhead: W69 Nuclear Warhead, 200 KT
Guidance: Inertial with terrain following radar
First Use Date: 1969
Producer: Boeing
Users: US Air Force
Other Designations: SRAM-A
Status: Was operational, withdrawn from use

The Short Range Attack Missile (SRAM) was a nuclear-armed air-to-surface weapon that replaced the AGM-28 Hound Dog on the B-52, and additionally armed the B-1B and FB-111 bombers. Normally carried on an eight missile rotary launcher in the bomb bay of the two larger bombers, it could also be carried on wing pylons (the only method of carriage on the FB-111). With a supersonic aircraft like the FB-111, tailcones were added to the AGM-69 to reduce drag in flight. Just after launch the tailcone was jettisoned from the missile before rocket motor ignition. AGM-69s were first assigned to the 42nd Bomb Wing at Loring, Maine in March of 1972, and went on alert for the first time on September 15, 1972. The SRAM was withdrawn from use by the Air Force in 1990 over safety concerns, and formally retired in 1993.

M-70 Unknown

Specifications

None

Unbuilt missile, purpose unknown, possible modification of the Minuteman ICBM to another role.

BGM-71 TOW

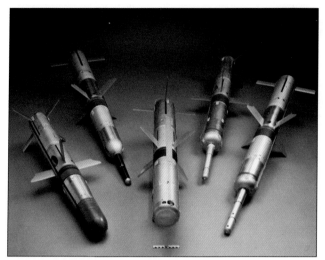

NEW GENERATION – The fly-over, shoot down TOW 2B (center), the latest version of the Hughes Aircraft Company's widely regarded BGM-71 TOW (Tube-launched, Optically-tracked, Wire-guided) antitank missile, is surrounded by other members of the TOW family. Other TOW versions, clockwise from left in order of introduction, are: the Basic TOW, the Improved TOW (ITOW) with a telescoping probe for standoff detonation, TOW 2 with heavier warhead, and the TOW 2A with tandem warheads. Photograph includes both display models and inert missiles.

MORE PUNCH – With its extendable probe thrust forward, a Hughes Aircraft Company BGM-71 TOW 2 anti-tank missile flies from the launch tube of the US Army's M2 Bradley Fighting Vehicle System during a test. TOW 2 features improved guidance and a more potent warhead designed to defeat advanced enemy armor. The family of TOW (Tube-launched, Optically-tracked, Wire-guided) missiles was developed for the Army Missile Command by Hughes Aircraft, a unit of GM Hughes Electronics.

Specifications

Length: 3 ft 10 in (116.84 cm)
Diameter: 6 in (15.23 cm)
Function: Air-to-Surface, Surface-to-Surface Anti-tank
Weight: 48 lbs (21 kg)
Warhead: 8–13 lbs High Explosive Shaped Charge
Guidance: Wire guided
First Use Date: 1965
Producer: Hughes
Users: US Army, US Marine Corps, Argentina, Bahrain, Botswana, Cameroon, Canada, Chile, Chad, Columbia, Denmark, Egypt, Ethiopia, Finland, Hungary, Germany, Greece, Iran, Israel, Italy, Japan, Jordan, Kenya, Kuwait, Lebanon, Luxembourg, Morocco, Norway, Oman, Pakistan, Portugal, Saudi Arabia, Somalia, Republic of Korea, Spain, Sweden, Swaziland, Switzerland, Taiwan, Thailand, Tunisia, Turkey, United Arab Emirates, United Kingdom, Vietnam, Yemen
Status: Operational

The Tube-launched, Optically-tracked Wire guided (TOW) missile was designed primarily for the anti-tank role, although it can be used against other ground targets, especially if they are hardened or armored. In the aerial role, the BGM-71 is carried by the UH-1, AH-1 or other helicopters, and in the ground role the TOW can be carried by a small vehicle or armored platform, Warheads on later versions, the BGM-71E/F, fired a shaped charge vertically downward into the tank's less-armored top as the missile flew just above it. However, these missiles retained the option to directly attack other kinds of targets.

MIM-72 Chaparral

M730 tracked vehicle and the MIM-72 Chaparral combined make up the M48 Chaparral weapons system. Left side view of the M730 with the MIM-72 turret turned to the left.

Right ½ front view of an M730 tracked vehicle firing an MIM-72 Chaparral surface-to-air missile.

Specifications

Length: 9 ft 5 in (287.02 cm)
Diameter: 5 in (12.7 cm)
Maximum Span: 2 ft ¾ in (62.86 cm)
Function: Surface-to-Air
Weight: 190 lbs (86 kg)
Warhead: 5 lbs High Explosive Fragmentation
Guidance: Infrared
First Use Date: 1966
Producer: Philco (Ford Aerospace)
Users: US Army, Chile, Columbia, Ecuador, Egypt, Israel, Morocco, Portugal, Taiwan
Status: Operational
Family: Sidewinder

For surface-to-air defense for troops in the field, the US Army looked at the AIM-9 Sidewinder as the basis for a defensive system following the cancellation of the MIM-46 Mauler. The Chaparral served with the US Army from the 1970s into the 1990s when it was replaced by more capable systems. Limitations with the MIM-72 were such that a target had to be visually tracked before the infrared seekers would acquire a heat source (normally engine exhaust of a departing aircraft). Improvements over the time in use negated the shortcomings of the Chaparral system. The United States' last operators of the MIM-72 were Army National Guard units.

Right ½ rear view of an MIM-72 Chaparral surface-to-air missile as it is fired from a trailer-mounted mobile launcher.

UGM-73 Poseidon

The 81st launch of a UGM-73 Poseidon C-3 was made from the USS *Will Rogers* at 12:45 p.m. EDT on May 20, 1986. This was also the 61st launch of a UGM-73 Poseidon from a submarine.

A UGM-73 Poseidon missile launch from aboard a ship. February 4th, 1970.

Specifications

Length: 34 ft (1036.32 cm)
Diameter: 6 ft 2 in (187.96 cm)
Function: Underwater-to-Surface sea launched ballistic missile
Weight: 65,000 lbs (29,510 kg)
Warhead: (10) W68 Nuclear Warhead, 50 KT or (14) W76 Nuclear Warhead, 100 KT
Guidance: Inertial
First Use Date: 1968
Producer: Lockheed
Users: US Navy
Other Designations: C-3
Status: Was operational, withdrawn from use

First launched in August of 1968, the UGM-73 gained its sea legs aboard a submarine in August 1970. As with other American SLBMs, the UGM-73 was cold launched, it was ejected from the submarine tube by gas and the solid rocket motors ignited when the missile broached the water's surface. The Poseidon was capable of carrying up to 14 multiple independently targeted reentry vehicle (MIRV) warheads and officially entered service in March of 1971, It was superseded by the UGM-96 Trident missile and withdrawn from use (and the submarines that launched it decommissioned) by September 1992.

MQM-74 / BQM-74 Chukar

Left ½ front view of a BQM-74C Chukar target missile on a handling dolly. Technician is grasping the rear launch lug atop the fuselage.

Left ½ rear view, from below, of a BQM-74C Chukar target missile mounted on the wing pylon of a Lockheed C-130 Hercules.

Specifications (BQM-74C)

Length: 12 ft 11½ in (394.96 cm)
Diameter: 1 ft 2 in (35.56 cm)
Maximum Span: 5 ft 9¼ in (175.89 cm)
Function: Aerial Target
Weight: 514 lbs (233 kg)
Guidance: Radio Controlled or Programmed
First Use Date: 1965
Producer: Northrop
Users: US Navy, US Air Force
Status: Operational

Primarily used in the aerial target role to simulate anti-ship missiles for naval defense system personnel, the MQM-74 can also act as a decoy when used against enemy forces while simulating cruise missiles of attack aircraft such as the Su-24 or Su-27. The Chukar (named for a type of Western American partridge) could also be outfitted with sensors and used in the reconnaissance role. Later versions added an air-launch capability and were designated as BQM-74 to reflect this new status. A later version of the Chukar, the BQM-74F, has swept-back tail and wings and a longer fuselage, making it look slightly different than the earlier versions of the BQM-74. Powerplant and other upgrades make the BQM-74F the highest performing member of the Chukar family.

M-75 Unknown

Specifications:

Dimensions: Unknown
Details: Unknown
Unbuilt

AGM-76 Falcon

Right side view of an AGM-76 under the wing of a US Air Force F-4 Phantom II.

Right ½ front view of an AGM-76 on the inboard pylon of a US Navy A-6 Intruder.

Specifications

Length: 13 ft 5 in (408.94 cm)
Diameter: 1 ft 1½ in (34.29 cm)
Maximum Span: 2 ft 9 in (83.82 cm)
Function: Air-to-Surface
Weight: 225 lbs (102 kg)
Warhead: High Explosive
Guidance: Radar Homing
First Use Date: 1966
Producer: Hughes
Users: US Air Force, US Navy
Status: Tested

A further variant of the Falcon missile family was the AGM-76 missile. The AIM-47 air-to-air missile was to be converted into a fast, long-range missile called AGM-76 to destroy enemy surface-to-air missile (SAM) sites in North Vietnam before the launch aircraft came into the lethal zone of the SAM missile. This would be accomplished by the missile homing in on the enemy radar's emissions. But after the AGM-76 concept was approved by the Air Force, the US Navy promoted an air-to-ground version of their existing Standard missile developed for this purpose and the AGM-78 won. A test missile of the AGM-76 was donated to the Smithsonian Institution in 1970 by Hughes Aircraft.

FGM-77 Dragon

Left ½ front view as an anti-tank gunner fires a FGM-77 Dragon missile.

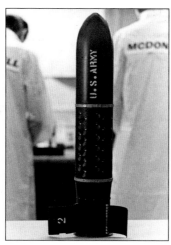

View of an FGM-77 Dragon missile standing upright on a checkout table. The folding fins are in the extended position.

BEIRUT, LEBANON – A US Marine of the 32nd Marine Amphibious Unit (MAU) carries his FGM-77 Dragon anti-tank/assault missile as he prepares to come ashore to assume management of the Port of Beirut.

Specifications

Length: 2 ft 5 in (73.66 cm)
Diameter: 9¾ in (24.76 cm)
Maximum Span: 13¼ in (33.65 cm)
Function: Surface-to-Surface Anti-tank
Weight: 24 lbs (11 kg)
Warhead: 5.5 lbs High Explosive Shaped Charge
Guidance: Wire Guided
First Use Date: 1973
Producer: McDonnell Douglas
Users: US Army, Jordan, Iran, Morocco, Netherlands, Saudi Arabia, Switzerland, Thailand, Yemen
Other Designations: MAW (Medium Anti-tank Weapon)
Status: Operational

Dragon is a man-portable anti-tank missile, fired from a disposable tube which also serves as the shipping container. A reusable tracking sight attaches to the tube, as can be a bipod/tripod as an aid in stabilizing during sighting. The FGM-77 is wire-guided for line of sight flight. The operator tracks the missile via a tail flare, while the missile propelled by a multitude of rocket motors along the sides at the rear of the missile. The Dragon was envisaged as a replacement for the 90mm recoilless rifle, yet man-portable and shoulder fired. Size and weight constraints notwithstanding, the FGM-77 was able to destroy most armored vehicles, such as tanks, that an infantryman might encounter. The Dragon was originally known as MAW or Medium Anti-tank Weapon.

AGM-78 Standard ARM

Left ½ front view of an AGM-78 Standard ARM (Anti-Radiation Missile) under the wing of a Republic F-105 Thunderchief. Anti-radiation missiles home in on and destroy radar sites.

Left side view of a McDonnell-Douglas F-4G Wild Weasel aircraft in flight, banking away from the camera. Mounted on the aircraft, from left wing outboard across are: AGM-88 HARM, AGM-65 Maverick, AN/ALQ-119, AGM-78 Standard ARM and AGM-45 Shrike missile. The AN/ALQ-119 is a jamming pod carried for counter-measures, the rest are air-to-ground missiles.

Specifications

Length: 15 ft (457.20 cm)
Diameter: 1 ft 1 inch (33.02 cm)
Maximum Span: 3 ft 3½ in (100.33 cm)
Function: Air-to-Surface
Weight: 1,400 lbs (635 kg)
Warhead: 220 lbs High Explosive Fragmentation
Guidance: Radar Homing
First Use Date: 1967
Producer: General Dynamics
Other Designations: Anti Radiation Missile
Users: US Navy, US Air Force
Status: Operational

The Standard Anti-Radiation Missile (ARM) guides to emissions from ground-based radar antennas and destroys them. The AGM-78 is based on the Standard Missile family, and initially used a seeker head developed from the AGM-45 Shrike missile. When a flight of aircraft entered hostile territory, they were usually illuminated by radar to allow opposing weapons, either guns or missiles to take aim. If these radar units were to be destroyed before they could acquire a lock on the approaching aircraft, it would be a great value to the flight. The Standard ARM guides on the transmitted radar energy, much as a moth drawn to a light on a dark night. Once the radar was destroyed, the gun emplacements or ground radar commanded anti-aircraft missiles were blind.

AGM-79 Blue Eye

Specifications

Dimensions: Unknown
Function: Air-to-Surface
Warhead: High explosive
Guidance: Television
Design date: 1960s
Producer: Martin Marietta
Users: US Air Force
Status: Tested

The AGM-79 Blue Eye was an abortive attempt to produce a television-guided version of the AGM-12 Bullpup, which was similar in size and shape. Although tested in 1968, it was cancelled in favor of the AGM-65 Maverick.

AGM-80 Viper

Specifications

Dimensions: Unknown
Function: Air-to-Surface
Warhead: High Explosive
Guidance: Infrared
Design date: 1960s
Producer: Chrysler
Users: US Air Force
Status: Tested

Viper was an infrared-guided version of AGM-12C Bullpup for use against ground targets. The AGM-80 program did not get beyond testing, and the program was closed down in the 1970s, cancelled at the same time as the AGM-79.

View of the launch of an AGM-80 Viper missile.

AQM-81 Firebolt

Specifications

Length: 17 ft (518.16 cm)
Diameter: 1 ft 1 inch (33.02 cm)
Height: 2 ft 2 in (66.04 cm)
Maximum Span: 3 ft 4 in
Function: Air launched supersonic target drone
Weight: 1,231 lbs (559 kg)
Guidance: Radio Control, Programmed
First Use Date: 1983
Producer: Teledyne Ryan
Users: US Air Force, US Navy
Other Designations: HAST, Teledyne Ryan Model 305, Sandpiper
Status: Tested

Front ¼ right view, from below, of the AQM-81 Firebolt target missile.

The AQM-81 was a high-speed aerial target that that used a hybrid rocket motor that combined solid fuel with a liquid oxidizer. In this case the oxidizer was inhibited red fuming nitric acid (IRFNA). Packing a high-performance engine into such a small package made for a target that flew at speeds over Mach 4. The Firebolt was recoverable, coming down after a mission by a two-stage parachute, to be snatched mid-air by a helicopter using MARS, the Mid Air Retrieval System. Despite successful testing which concluded in 1985, production contracts for the AQM-81 were not awarded.

AIM-82 Dogfight

Specifications:

Dimensions: unknown
Function: Air-to-air

No images available

The AIM-82 was a proposed USAF air-to-air missile to replace the AIM-9 Sidewinder for use on the F-15 Eagle. Project ceased as the US Navy was developing a replacement (the AIM-95), which also did not enter production. AIM-82 did not proceed to a hardware stage, so no dimensions are given. "Dogfight" may be an unofficial nickname to show the purpose of the missile.

AGM-83 Bulldog

Left side view of a Douglas A-4E Skyhawk attack aircraft with an AGM-83 Bulldog missile mounted on the left outboard weapons pylon. Lt.Col. Bennie W. "Bulldog" Summers stands alongside, Naval Weapons Center, China Lake, California 19 January 1971.

Right side view, from below, of a Douglas TA-4F Skyhawk attack aircraft in flight with an AGM-83 Bulldog missile mounted on the right outboard weapons pylon.

Specifications

Length: 10 ft 6 in (320.04 cm)
Diameter: 1 ft (30.48 cm)
Maximum Span: 3 ft 1 inch (93.98 cm)
Function: Air-to-Surface
Weight: 620 lbs (281 kg)
Warhead: 250 lbs High Explosive Fragmentation
Guidance: Laser Guided
First Use Date: 1970
Producer: Texas Instruments, Naval Weapons Center
Users: US Navy
Status: Tested

Right ½ front view a Douglas A-4M Skyhawk attack aircraft with an AGM-83 Bulldog missile mounted on the right outboard weapons pylon.

The AGM-83 was an outgrowth of the AGM-12 Bullpup. The Bulldog removed the labor-intensive pilot commands of the Bullpup by replacing the radio control guidance suite with a laser-seeking head on the missile. The intended target would be illuminated from a separate laser source, such as one used by ground troops, and the Bulldog would then home in on the target. Some reports indicate that procurement of the AGM-83 may have taken place while waiting for a laser-guided version of the AGM-65 Maverick, but the Bulldog never received a production contract, and the Maverick did see multi-service use by the United States.

AGM-84 Harpoon

Right ½ front view of an ATM-84A Harpoon missile mounted on the outboard weapons pylon of a Lockheed P-3 Orion anti-submarine patrol aircraft. The ATM-84 is a training version of the AGM-84 Harpoon.

Right ½ front view of a sharply banked Lockheed P-3 Orion anti-submarine patrol aircraft in flight. The P-3 is carrying an AIM-9 Sidewinder missile on each outboard pylon, and four AGM-84 Harpoon missiles mounted on the wings and under the fuselage.

Specifications

Length: 12 ft 7 in (383.53 cm)
Diameter: 1 ft 1 inch (33.02 cm)
Function: Air-to-Surface, Surface-to-Surface Anti-shipping
Weight: 1,160 lbs (526 kg)
Warhead: 488 lbs High Explosive Fragmentation
Guidance: Radar
First Use Date: 1972
Producer: McDonnell-Douglas
Users: US Navy, Australia, Brazil, Canada, Chile, Denmark, Egypt, Germany, Greece, Iran, Israel, Indonesia, India, Japan, Malaysia, Netherlands, Pakistan, Poland, Portugal, Saudi Arabia, Singapore, Republic of Korea, Spain, Taiwan, Thailand, Turkey, United Arab Emirates, United Kingdom
Other Designations: UGM-84, RGM-84
Status: Operational

The Harpoon is an air-to-surface or surface-to-surface anti-shipping missile. The AGM-84 can be carried in US service by the P-3 Orion, S-3 Viking, A-6E, F/A-18A-F or B-52G/M; the UUM-84 is fired from a submarine, while the RGM-84 variant is fired from ship-mounted launchers. The UUM/RGM-84 is somewhat longer, being equipped with short boosters. to elevate the Harpoon to a ballistic trajectory before the air-breathing turbine engine directs the missile to the target.

RIM-85

Specifications:

Dimensions: Unknown

No images available

The RIM-85 was to have been a ship-launched medium range all-weather missile for use by the US Navy against air or surface targets, but was cancelled before work proceeded past the concept stage.

AGM-86 ALCM

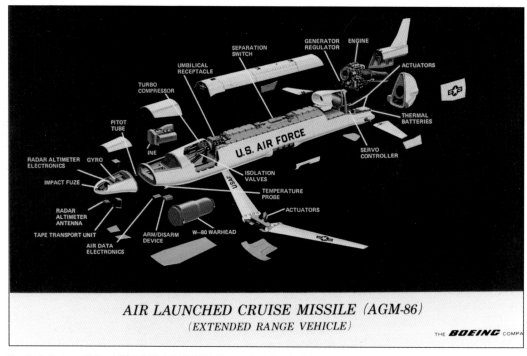

Exploded view of the AGM-86B ALCM (Air Launched Cruise Missile).

Specifications

Length: 20 ft 9 in (632.46 cm)
Diameter: 24.5 in (62.23 cm)
Maximum Span: 12 ft (365.76 cm)
Function: Air-to-surface Strategic Cruise Missile
Weight: 3, 150 lbs (1,430 kg)
Warhead: 1,000-3,000 pound High Explosive or W80 Nuclear Warhead, 200 KT
Guidance: Inertial Navigation
First Use: 1982
Producer: Boeing
Users: US Air Force
Status: Operational

Left ½ front view of the AGM-86B ALCM (Air Launched Cruise Missile) at left and the shorter AGM-86A on the right.

The Air Launched Cruise Missile (ALCM) was originally developed as a decoy, but evolved into a standoff weapon capable of carrying nuclear or conventional warheads. Arms control treaties have limited their replacement or the upgrading of their nuclear capabilities, but the AGM-86 has found new purpose with revised conventional warheads. While projected for use on several aircraft, thus far the B-52 Stratofortress has been the only aircraft to carry the AGM-86 in combat.

AGM-86B ALCM (Air Launched Cruise Missile) in flight over desert in the American West during the missile test program.

Left side view of the AGM-86B ALCM (Air Launched Cruise Missile) as it is loaded onto a wing pylon of a Boeing B-52 Stratofortress at Griffiss AFB, New York.

AGM-87 Focus

Specifications

Length: 9 ft 5 in (287.02 cm)
Diameter: 5 in (12.7 cm)
Function: Air-to-Surface
Weight: 155 lbs (70 kg)
Warhead: High Explosive
Guidance: Infrared
First Use Date: 1960s
Producer: Naval Weapons Center, General Electric
Users: US Navy
Status: Possibly tested

The AGM-87 is purported to be a derivative of the AIM-9 Sidewinder missile, with the avionics optimized to use the infrared seeker head to locate ground targets producing a heat signature, such as trucks. Reportedly used in Southeast Asia during 1969–70, some records show the program to still be classified.

AGM-88 HARM

Right ½ front view of the AGM-88 HARM (High-speed Anti-Radiation Missile) mounted on a display.

View from a remote camera as an AGM-88 HARM is about to strike a target ship during testing.

Specifications

Length: 13 ft 8 in (416.56 cm)
Diameter: 10 in (25.4 cm)
Maximum Span: 3 ft 8 in (111.75 cm)
Function: Air-to-Surface Anti-Radiation Missile
Weight: 800 lbs (363 kg)
Warhead: 145 lbs High Explosive Fragmentation
Guidance: Radar Homing
First Use: 1983 (Navy) 1984 (Air Force)
Producer: Raytheon, Texas Instrument, Ford Aerospace, Naval Weapons Center
Users: US Air Force, US Navy
Status: Operational

Left side view of an General Dynamics F-16B Fighting Falcon in flight with an AGM-88 under the left wing, an AIM-9 Sidewinder mounted on the wingtip rail.

The AGM-88 is a High-speed Anti-Radiation Missile (HARM) designed to detect and destroy enemy radar operating as air defense systems. The radar seeker in the HARM homes in on the radar emissions of the defensive positions antennae and destroys the antenna and the surrounding area, which often includes the command and control emplacements for the defensive systems. The AGM-88 was first used on March 24, 1986 by Navy A-7Es against Libya.

UGM-89

Specifications:

Dimensions: Unknown
Unbuilt

Proposed submarine-launched anti-submarine/anti-ship missile. Program cancelled before construction.

BQM-90

Specifications:

Dimensions: Unknown
Unbuilt

Remotely-controlled, high-altitude, supersonic aerial target for the US Navy, the BQM-90 program was canceled before construction.

AQM-91 Firefly

Right side view from below of a Lockheed DC-130 Hercules drone control-launch aircraft carrying a load of two AQM-91 Compass Arrow drones.

Left side view of an AQM-91 Compass Arrow drone suspended underneath a recovery helicopter, a Silkorsky HH-3 Jolly Green Giant. The drone is steadied underneath the helicopter by the small stabilization parachute. After a drone mission, the AQM-91 is recovered by parachute, picked up by helicopter and flown back to the support base.

Specifications

Length: 34 ft 2 in (1041.39 cm)
Height: 3 ft 4 in (101.6 cm)
Maximum Span: 47 ft 8 in (1452.87 cm)
Function: Aerial Reconnaissance
Weight: 5,400 lbs (2,451 kg)
Guidance: Programmed

First Use Date: 1968
Producer: Teledyne Ryan
Users: US Air Force
Other Designations: Ryan Model 154, Compass Arrow
Status: Tested

Developed to counter the inability of traditional manned reconnaissance aircraft to photograph high-threat areas of Asia and Southeast Asia during the Vietnam War, the AQM-91 had the task of high-altitude, long-distance flight once launched. After launch from a DC-130 Hercules mothership, the Firefly drone would enter an area of interest at the selected altitude, and return, to be recovered by helicopter as it descended by parachute. Once on the ground the film payload would be processed or sensors would be downloaded for intelligence. After a test program that became less secret when one craft landed in public view in 1969, the program was shelved in July 1973 as relations between the United States and China improved. When the need arose again for image in the region, imaging satellite technology had improved, which was more discreet than reintroducing the AQM-91 program.

FIM-92 Stinger

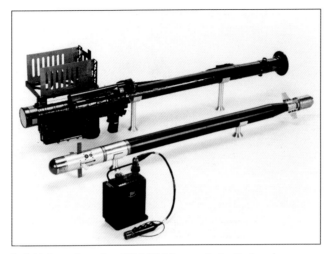

Left ½ front view of an FIM-92 Stinger missile displayed alongside the shoulder launcher.

Left ½ front view of a US Army soldier firing an FIM-92 Stinger missile from the shoulder launcher.

Specifications

Length: 5 ft (152.4 cm)
Diameter: 2¾ in (6.98 cm)
Function: Surface-to-air missile
Weight: 35 lbs (16 kg)
Warhead: 6.5 lbs High Explosive Fragmentation
Guidance: Infrared
First Use Date: 1975
Producer: Hughes, Raytheon
Users: US Army, US Navy, US Marine Corps, Afghanistan, Angola, Austria, Bahrain, Chad, Denmark, Egypt, France, Germany, Greece, Iran, Israel, Italy, Japan, Kuwait, Netherlands, Pakistan, Qatar, Saudi Arabia, South Korea, Switzerland, Taiwan, Turkey
Status: Operational

A man-portable surface-to-air missile, the FIM-92 is known throughout the world as the Stinger. Designed to provide a measure of air defense for the individual soldier, the speedy Mach 2 Stinger has been adapted for use as an air-to-air weapon on helicopters, as well as multiple weapon mounts for ships and vehicles. Over 70,000 of the missiles have been produced. The FIM-92 started replacing the FIM-43 Redeye in US use in the late 1970s and is still operational. The Stinger is fired from a disposable tube, to which is attached a reusable trigger, sight and power supply. The missile leaves the tube via a gas charge, and the rocket motor is ignited when the Stinger is a few yards from the operator. The mobile version which mounts the Stinger on the HMMWV (Hummer) chassis is known as the M1097 Avenger. Stingers supplied to Afghan rebels were highly successful against Soviet aircraft in the Afghan War.

GQM-93

Specifications

Length: 29 ft 7 in (901.69 cm)
Diameter: 10 ft 8 in (325.12 cm)
Maximum Span: 57 ft (1737.36 cm)
Function: Reconnaissance Drone
Weight: 5,300 lbs (2,406 kg)
Guidance: Radio Control
First Use Date: 1970
Producer: E-Systems
Users: US Air Force
Other Designations: L450F, Compass Dwell, XQM-93, YQM-93
Status: Tested

The GQM-93, shown in the piloted variant, while in storage after completion of testing. The NOAA titles appear to be spurious, as NOAA Aircraft Operations has no record of the craft.

The GQM-93 was a reconnaissance drone based largely on the Schweizer SGS 2-32 sailplane. Intended for use for high-altitude and long-endurance surveillance, the GQM-93 was tested but not adopted by the US Air Force (and neither was its competitor, the Martin Marietta Model 845A). While in flight testing, the first of two of the drones crashed, but the second completed the evaluation program. The NOAA titles shown on the second vehicle while in storage would seem to denote that the aircraft was operated by the National Oceanic and Atmospheric Administration. NOAA's Aircraft Operations section has no record of the GQM-93, but after completing evaluation by the Air Force, it may have been shown to some NOAA staff by E-Systems as a possible aerial platform for their scientific work and collection of ocean data.

GQM-94 Compass Cope B

Right ½ front view of a GQM-94 Compass Cope B drone as it is prepared for flight.

View from directly below of a GQM-94 Compass Cope B drone in flight.

Specifications

Length: 40 ft (1219.20 cm)
Height: 12 ft 8 in (386.08 cm)
Maximum Span: 90 ft (2743.20 cm)
Function: Reconnaissance Drone
Weight: 14,400 lbs (6,537 kg)
Guidance: Radio Control
First Use Date: 1973
Producer: Boeing
Users: US Air Force
Designations: Compass Cope B, B-Gull, YGQM-94
Status: Tested

Left ¼ rear view of a GQM-94 Compass Cope B drone in flight, trailed by chase aircraft.

The GQM-94 was a high-altitude, long-duration aerial reconnaissance drone with projected photographic and electronic surveillance capabilities. Two drones were built, the first crashing shortly after its first flight. Competing with the GQM-98 built for the same requirements, the "Compass Cope" series did not go beyond testing, and the project was terminated. The surviving GQM-94 was retired to the National Museum of the United States Air Force in September 1979.

AIM-95 Agile

View of the test firing of an AIM-95 missile.

Left side view of a McDonnell F-4 Phantom II parked on a ramp with an AIM-95 Agile missile mounted on the left wing pylon.

Specifications

Length: 8 ft (243.84 cm)
Diameter: 8 in (20.32 cm)
Function: Air-to-Air
Warhead: High Explosive
Guidance: Infrared
Design Date: 1971
Producer: Naval Weapons Center
Users: US Navy, US Air Force
Status: Tested

For use on the F-14 Tomcat, the Naval Weapons Center, China Lake, California developed the AIM-95 to replace the AIM-9 Sidewinder, itself developed at China Lake. Agility in the Agile was provided by a gimbaled engine for thrust vectoring, and the infrared seeker head was optimized for wide aspect tracking of opposing aircraft. But like many of the hoped for replacements for the Sidewinder and AIM-7 Sparrow, the Agile as the AIM-95 was known, was not to succeed and never went into production.

Lt. Carl Cline and XAIM-95 Agile missile, at China Lake, June 19, 1970.

UGM-96 Trident I

The US Navy provisions its strategic missile submarines at coastal bases in the states of Washington, South Carolina and Georgia. The new *Ohio*-class submarines are designed to carry the 44-ft-long and 83-in-diameter UGM-133 Trident II missiles when deliveries begin in 1989. Until then, the submarines so far introduced into service are adapted to carry the smaller (34 ft long and 74 in in diameter) UGM-96 Trident I.

The tenth UGM-96 Trident I test launch lifted off successfully at 3:31 p.m. EST on January 17, 1978 from Cape Canaveral Air Force Station, Florida.

Specifications

Length: 34 ft (1036.32 cm)
Diameter: 6 ft 2 in (187.96 cm)
Function: Underwater-to-Surface sea launched ballistic missile
Weight: 73,000 lbs (33,142 kg)
Warhead: (8) W76 Nuclear Warhead, 100 KT
Guidance: Inertial
First Use Date: 1977
Producer: Lockheed
Users: US Navy
Other Designations: C-4
Status: Was operational, withdrawn from use

With improvements to weapons technology for warhead, guidance and propulsion, the Poseidon missile was due for replacement, and that came with the Trident. Designated the UGM-96 for the Trident I, the missile entered service with the *Ohio*-class Ballistic Missile Submarine fleet in 1979. The UGM-96 could carry eight nuclear warheads in its rounded nose, which was topped by an aerospike. The spike would extend after the missile was cleared from the water. In use for about twenty years, the UGM-96 has been replaced on the US Navy *Ohio*-class submarines by the improved UGM-133 Trident II.

The 21st demonstration and shakedown operational launch of a UGM-96 Trident I missile from the USS *Henry M. Jackson* and the 46th flight of the Trident I took place at 2:54 p.m. EST, December 4, 1984.

AIM-97 Seekbat

Specifications

Length: 15 ft (457.2 cm)
Diameter: 1 ft 1½ in (34.29 cm)
Maximum Span: 3 ft 6½ in (107.94 cm)
Function: Air-to-Air
Weight: 1,300 lbs (590 kg)
Warhead: 220 lbs, High Explosive Fragmentation
Guidance: Infrared
Design Date: 1972
Producer: General Dynamics
Users: US Air Force
Status: Tested

Right ½ front view of an AIM-97 Seekbat missile mounted on a display stand in a hallway at the Air Force Armament Lab at Eglin AFB, Florida.

The AIM-97 was designed to be launched from F-4Es to counter the MiG-25 Foxbat until the F-15 became operational. It used the airframe of the AGM-78 Standard ARM, with a larger propulsion unit and an infrared seeker head to home in on the exhaust produced by the MiGs large Tumansky turbojet engines. By the late 1970s the Seekbat missile had been cancelled, due to tighter budgets and the reassessment of the threat posed by the Foxbat aircraft. Six of the Seekbat missiles were available for use during the 1973 Yom Kippur War in case US forces would have been used in that conflict.

GQM-98 Compass Cope

Right ½ rear view, from slightly below, of a GQM-98 Compass Cope drone in flight, shortly after takeoff.

Front view, from slightly above, of the two GQM-98 Compass Cope drones undergoing checkout in a hangar.

Specifications

Length: 37 ft (1127.76 cm)
Height: 8 ft (243.84 cm)
Maximum Span: 81 ft (2468.88cm)
Function: Reconnaissance Drone
Weight: 14,000 lbs (6,356 kg)
Guidance: Radio Control
First Use Date: 1974
Producer: Teledyne Ryan
Users: US Air Force
Other Designations: Ryan Model 235, Compass Cope, R-Tern
Status: Tested

The Compass Cope was the Ryan competitor for a high-altitude, long-endurance reconnaissance drone, built to compete against the GQM-94. Similar in layout, each of the craft was large for an unmanned drone, the slightly smaller GQM-98 having a wingspan of 81 ft. The GQM-98 is also notable for setting a record for unmanned, unrefueled flight on one of its test missions, staying airborne for 28 hours and 11 minutes. The Compass Cope R flew until September 1975 when the demonstration and evaluation studies concluded, but no follow-on contracts for a production vehicle based on either craft were awarded. The program was terminated in 1977.

LIM-99

Specifications: None

Project cancelled: No known images of this missile

LIM-100

Specifications: None

Project cancelled: No known images of this missile

RIM-101

Specifications: None

Project cancelled before progression beyond designation.

No known images of this missile.

PQM-102 Delta Dagger

Left ¼ front view of a PQM-102B Delta Dagger aircraft now used as a aerial target drone. The Delta Dagger was designated as the F-102A when used as a fighter by the US Air Force.

Left ½ rear view of a Convair F-102A Delta Dagger in flight. Delta Daggers were converted when they became obsolete into aerial targets. Early unmanned Daggers were PQM-102A models with large radio gear taking the place of the pilot and ejection seat. When the radio gear became smaller allowing instances where a pilot could fly the aircraft, they were designated PQM-102B models.

Specifications

Length: 68 ft 5 in (2085.33 cm)
Height: 21 ft 2½ in (646.33 cm)
Maximum Span: 38 ft 1½ in (1162.05 cm)
Function: Aerial Target
Weight: 24,500 lbs (11,123 kg)
Guidance: Radio Control
First Use Date: 1974
Producer: Sperry
Users: US Air Force
Other Designations: F-102A, QF-102A
Status: Was operational, withdrawn from use

When first converted from Convair F-102 Delta Dagger interceptor fighters, the craft had no provision for a pilot, and operated as NOLO (No Live Operator) target drones. Later conversions used avionics kits of reduced size, enabling a piloted drone if required by the circumstances. The Delta Dagger drones were used as targets by both the US Air Force and US Army, with the PQM-102 being on the receiving end of air-to-air as well as surface-to-air missiles. The Delta Dagger at one point was heavily targeted by its stable mate the F-106, but as stocks of excess -102s ended, the now surplus Delta Darts then became QF-106 targets.

AQM-103

Artist's view of an AQM-103 drone in flight.

Left front view of an XAQM-103 drone mounted on the outboard wing pylon of a DC-130 Hercules.

Specifications

Length: 29 ft (883.92 cm)
Maximum Span: 27 ft (822.96 cm)
Function: Aerial Reconnaissance
Weight: 3,000 lbs (1,362 kg)
Guidance: Television
First Use Date: 1975
Producer: Teledyne Ryan
Users: US Air Force
Other Designations: XQM-103
Status: Tested

The AQM-103 was an extensively modified Teledyne Ryan Model 147G Firebee reconnaissance drone with the structure strengthened and optimized for highly maneuverable flight, the airframe beefed up to withstand a 10 G load in-flight. This "high g" capability could either have been used to enhance survivability in a heavily defended target area, or to use the drone itself in an air-to-air role. The AQM-103 was air-launched from a DC-130 Hercules, and recovered by helicopter in mid-air as it descended under a parachute. Testing in the mid-1970s came to an end quickly, although the slightly modified airframe was spotted on display at an air show in 1979, but apparently not as part of a viable program.

MIM-104 Patriot

Left ½ rear view of a truck-mounted MIM-104 Patriot missile launcher. There are three containers mounted on this launcher, normally four would be carried.

Left side view as an MIM-104 Patriot missile is launched from a truck-mounted missile launcher.

Specifications

Length: 17 ft 5 in (530.86 cm)
Diameter: 1 ft 4 in (40.64 cm)
Maximum Span: 2 ft 9 in (83.82 cm)
Function: Surface-to-Air
Weight: 1,980 lbs (899 kg)
Warhead: 150 lbs, High Explosive Fragmentation
Guidance: Radar
First Use Date: 1973
Producer: Raytheon
Users: US Army, Germany, Greece, Israel, Japan, Kuwait, Netherlands, Saudi Arabia, Taiwan
Other Designations: SAM-D
Status: Operational

Right ½ front view as an MIM-104 Patriot missile is launched from a truck-mounted missile launcher. Note the missile is fired from the upper left container; the lower right missile has already been fired.

The MIM-104 is an air-defense missile used as protection from aircraft, cruise missiles and ballistic missiles on the battlefield. The size of the Patriot allows for carriage of up to four missiles in a launcher. The Patriot has been upgraded; these missiles are called Patriot Advanced Capability (PAC) -1 and PAC-2. The PAC-3 weapon is a total redesign and does not fall under the MIM-104 designation, but as yet has not received a separate MDS number.

The PAC-2 version of the missile apparently achieved success in Operation Desert Storm in 1991 by shooting down Iraqi ballistic SCUD missiles, but there are no confirmed kills.

MQM-105 Aquila

Left side view of an MQM-105 Aquila field team recovering a drone in the net of the rearmost vehicle, while the truck in front has an MQM-105 drone ready to be launched.

Right side view of an MQM-105 Aquila as it is recovered.

Specifications

Length: 6 ft (182.88 cm)
Diameter: 2 ft 6 in (76.2 cm)
Maximum Span: 11 ft 11 in (363.22 cm)
Function: Reconnaissance Drone
Weight: 330 lbs (150 kg)
Guidance: Radio/Television Controlled
First Use Date: 1975
Producer: Lockheed
Users: US Army
Other Designations: TADAR (Target Acquisition Designation and Aerial Reconnaissance)
Status: Tested

Left ½ rear view, from above, of an MQM-105 Aquila drone in flight.

The MQM-105 was a proposed battlefield reconnaissance craft, flown via remote control. After catapult launch, the Aquila would scout the near battlefield for targets, and could relay a television image back for assessment. If a target was verified, the MQM-105 could use a laser designator which would then enable laser-guided weapons to destroy the target. The loitering Aquila could then provide damage assessment of the target area and provide confirmation of the target's destruction.

BQM-106 Teleplane

The BQM-106 and the support vehicles.

Left side view of the BQM-106 in flight.

The BQM-106 was landed directly on a keel skid under the airframe.

Size of the craft made refurbishment between flights easier than many other drones.

Specifications

Length: 10 ft (304.8 cm)
Height: 2 ft 9 in (83.82 cm)
Maximum Span: 12 ft (365.76 cm)
Function: Aerial Drone Testbed
Weight: 235 lbs (106 kg)
Guidance: Radio Control
First Use or Design Date: 1975
Producer: US Air Force Flight Dynamics Laboratory
Users: US Air Force
Status: Tested

The BQM-106 was an experimental drone used to research and test capabilities, mechanisms and techniques that could be applied to future aerial drones. Various methods of flight control, airframe construction and payloads were tested in these vehicles produced by the US Air Force Flight Dynamics Laboratory at Wright-Patterson AFB in Ohio. There were several configurations tested of the Teleplane drone, the dimensions varied between these vehicles.

MQM-107 Streaker

Left ½ front view of an MQM-107 Streaker drone at launch from Launch Complex 32 at White Sands Missile Range, New Mexico.

Right ½ rear view of an MQM-107 Streaker drone on launch platform.

Specifications

Length: 18 ft 1 inch (551.18 cm)
Diameter: 1 ft 3 in (38.1 cm)
Maximum Span: 9 ft 10.5 in (300.99 cm)
Function: Aerial Target
Weight: 1067 lbs (484 kg)
Guidance: Ground Controlled or Pre-programmed
First Use Date: 1973
Producer: Beech Aircraft, Raytheon
Users: US Air Force, US Navy, US Army
Other Designations: Beech 1089, VSTT (Variable Speed Training Target), TEDS (Tactical Expendable Drone System)
Status: Was operational, withdrawn from use

The MQM-107 target drone is normally used as a high-speed target tug, as in the course of action it will fly a "racetrack" pattern towing a target sleeve, radar-augmented targets or infrared targets. But the Streaker may also be the target itself, flying at low altitudes to simulate a cruise missile or other low-flying threat. The MQM-107 is recoverable by parachute, and depending on if a water or hard surface landing is envisaged, a shock absorbing nose can be mounted on the airframe. Normally ground launched, the Streaker was tested for air launch from platforms such as the Lockheed DC-130 or McDonnell-Douglas F-4.

BQM-108

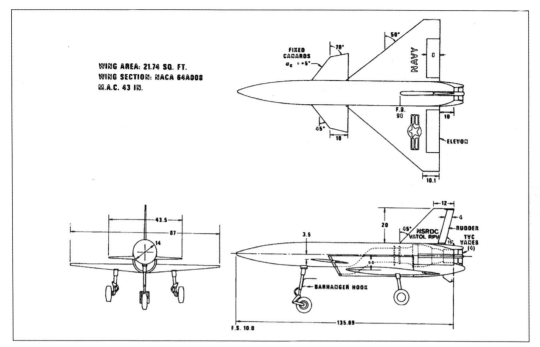

Three-view drawing of the BQM-108.

Specifications

Length: 11 ft 9 in (358.14 cm)
Diameter: 1 ft 2 in (53.34 cm)
Maximum Span: 7 ft 3 in (220.98 cm)
Function: Concept Testbed
Weight: 448 lbs (204 kg)
Guidance: Radio Control
First Use Date: 1976
Producer: David Taylor Naval Ship Research and Development Center (US Navy)
Users: US Navy
Other Designations: VATOL (Vertical Attitude Take Off and Landing)
Status: Tested

Ingenuity from engineers at the David W. Taylor Naval Ship Research and Development Center at Carderock, Maryland just outside Washington, DC brought forth the XBQM-108A. A technology testbed, it likely should have been given an "X-for-research" designation rather than an "M-for-missile" designation, but it did get the job done as an example to study a vertical attitude takeoff and landing (VATOL) vehicle.

Composed of parts from a MQM-74A drone, guidance and related components of an AGM-84 Harpoon missile and the landing gear from the diminutive single pilot Bede BD-5J Microjet, the BQM-108 was successfully flight tested in a hover mode while tethered in late summer 1976.

AGM/BGM/RGM/UGM-109 Tomahawk

Production supervisor Duane L. Daugherty inspects BGM-109 Tomahawk cruise missiles that have reached the end of the production line at General Dynamics Convair Division in San Diego, California. The missiles will receive paint touch-ups and will be stored in capsules for shipment to the US Navy.

Left ½ rear view as a BGM-109 Tomahawk cruise missile breaks the surface of the water after an underwater launch.

Explosion and fireball from a 1,000-lb warhead of a UGM-109 Tomahawk cruise missile. The Tomahawk was launched from a submarine submerged off the southern California coast, and crossed to a warehouse-sized target on a Navy range on San Clemente Island.

A UGM-109 Tomahawk cruise missile approaches overhead its target, a parked aircraft. The Tomahawk was launched from a submarine submerged off the southern California coast, and crossed to the revetted target on a Navy range on San Clemente Island.

A UGM-109 Tomahawk cruise missiles 1,000-lb warhead explodes in a fireball at the prescribed altitude over the target, a parked aircraft. The Tomahawk was launched from a submarine submerged off the southern California coast, and crossed to the revetted target on a Navy range on San Clemente Island.

The explosion and blast fragments from the UGM-109 Tomahawk cruise missile's 1,000 lb warhead destroy the target, a parked aircraft.

Specifications

Length: 18 ft 3 in (556.26 cm)
Diameter: 1 ft 9 in (53.34 cm)
Maximum Span: 8 ft 9 in (266.7 cm)
Function: Surface-to-Surface, Air-to-Surface
Weight: 2,650 lbs (1,203 kg)
Warhead: 1000 lbs, High Explosive or W80 Nuclear Warhead, 200 KT
Guidance: Inertial and Terrain following
First Use Date: 1976
Producer: General Dynamics, Hughes
Users: US Navy, US Air Force
Other Designations: SLCM, GLCM, Gryphon
Status: Operational

The debris remaining after the explosion and blast fragments from the UGM-109 Tomahawk cruise missile's 1,000 lb warhead which destroyed a parked aircraft.

The Tomahawk cruise missile is a multi-mode missile, and can found launched from aircraft, from mobile launchers, and from surface vessels as well as submarines. The BGM-109 is named the Gryphon, but seems to be popularly called the Tomahawk like the rest of the series. The term GLCM also applies, as it is the Ground Launched Cruise Missile. The SLCM or Sea Launched Cruise Missile is the more numerous and most well known of the M-109 family, either as the submarine-launched UGM-109 or ship launched RGM-109. Although initially fielded with a nuclear capability, the Tomahawk has been most effective with conventional warheads, or carrying submunitions. During the 1991 Gulf War, the submunitions included spools of carbon fiber to short out electrical power lines.

BGM-110

Right ½ rear view of a ZBGM-110 Cruise Missile on a support dolly. The Z prefix denotes a test vehicle of the BGM-110. After two flights, the BGM-110 lost a competition for the Cruise Missile to the BGM-109 Tomahawk.

Specifications

Length: 17 ft 10 in (543.56 cm)
Diameter: 1 ft 9 in (53.34 cm)
Maximum Span: 10 ft 6 in (320.04 cm)
Function: Air-to-Surface
Weight: 2,500 lbs (1,135 kg)
Warhead: W80 Nuclear Warhead, 200 KT
Guidance: Inertial and Terrain following
First Use Date: 1976
Producer: Vought
Users: US Navy
Status: Tested

The BGM-110 was an entry into the fly-off competition for an air-launched cruise missile. After a small number of flight tests, the contract was awarded to the BGM-110's competitor, the BGM-109, and further work was halted on this Vought proposal.

BQM-111 Firebrand

Pacific Missile Test Center, Point Mugu, California. A BQM-111 Firebrand Missile Target parked at the ground launch complex. Right side view.

View of various drones made by Ryan Aeronautical in San Diego. Front to rear, the AQM-81 Firebolt, the BQM-111 Firebrand, the BQM-34F Firebee II, and the BQM-34S Firebee.

Specifications

Length: 34 ft (1036.32 cm)
Diameter: 2 ft 4 in (71.12 cm)
Maximum Span: 9 ft (274.32 cm)
Function: Aerial target
Weight: 6,000 lbs (2,724 kg)
Guidance: programmed
Design Date: 1977
Producer: Teledyne Ryan
Users: US Navy
Status: Built, not flight tested

Left ½ front view from above of the BQM-111 drone on the ground. A person stands alongside to show the large size of the vehicle.

Firebrand was intended to be a ramjet-powered supersonic aerial target for the US Navy, to mimic a threat from a high-speed anti-shipping missile. The BQM-111 was to be launched from either a ground launcher or from a DC-130 aircraft, and accelerate to the speeds needed to operate the ramjets by solid rocket motor boosters. The BQM-111 program was terminated before flight-testing was started.

AGM-112

Electro-optically guided glide bomb, more commonly known as the GBU-15.

The GBU-15 (AGM-112) as dropped from an F-4E Phantom II.

Specifications

Length: 12 ft 11 in (393.7 cm)
Diameter: 1 ft 6 in (45.72 cm)
Maximum Span: 5 ft (152.4 cm)
Function: air-to-surface
Weight: 2, 600 lbs (1,180 kg)
Warhead: 2,000 lbs, High Explosive
Guidance: Electro-optical
First Use Date: 1983
Producer: Rockwell
Users: US Air Force
Other Designations: GBU-15 (TV)
Status: Operational

The AGM-112 is an electro-optically guided glide bomb, and as it is unpowered, it is normally known under the more appropriate designation of GBU-15. The AGM-112 designation is unused. The weapon consists of a Mark 84 bomb casing with added fin units for control and an electro-optical seeker head for guidance from the targeting aircraft.

RIM-113

Specifications

None
Function: Surface-to-air
Program cancelled

The RIM-113 was to have been a surface-to-air anti-cruise missile for the US Navy.

AGM-114 Hellfire

View of technicians loading the 5,000th AGM-114 Hellfire missile into its shipping container.

Left ¼ front view of a Hughes AH-64 Apache helicopter armed with four AGM-114 Hellfire missiles under each inboard weapons pylon. Outboard pylons hold 19-shot 2.75-in rockets.

Left ½ rear view of a Hughes AH-64 Apache helicopter firing an AGM-114 Hellfire missile.

Specifications

Length: 5 ft 4 in (162.56 cm)
Diameter: 7 in (17.77 cm)
Maximum Span: 2 ft 4 in (71.12 cm)
Function: Air-to-Surface, Anti-armor
Weight: 105 lbs (47 kg)
Warhead: 16 lbs, High Explosive, Shaped Charge
Guidance: Laser Guided
First Use Date: 1978
Producer: Rockwell/Martin Marietta (now Boeing/Lockheed-Martin)
Users: US Army, US Navy, US Marine Corps, Greece, Israel, Norway, Singapore, Sweden, Taiwan
Other Designations: Brimstone
Status: Operational

The AGM-114 is a laser-guided anti-tank missile. Known as the Hellfire, it is an air-to-surface missile normally employed by helicopters of the US Army, Navy and Marine Corps. The Hellfire is a subsonic missile, and can be used as an air-to-air weapon against other helicopters, or against slower moving fixed-wing aircraft. The Hellfire does have an autopilot in case there is a lapse in the laser designation tracking in mid-flight.

Sweden and Norway use the Hellfire for coastal defense, and Israel is reported to use Hellfire as a shipboard missile on small vessels. The Brimstone variant is used by the Royal Air Force. This version is used air-to-surface from high-speed aircraft rather than helicopters. The latest version of the Hellfire is the AGM-114L Longbow, which uses a MM-wave radar seeker. The AH-64 helicopters that field this version of the missile are called Apache Longbows.

MIM-115 Roland

Right ½ front view of a truck-mounted MIM-115 Roland surface-to-air missile system.

Right ½ front view of an XM975 tracked vehicle-mounted MIM-115 Roland surface-to-air missile system firing a missile.

Specifications

Length: 7 ft 11 in (241.3 cm)
Diameter: 6 in (15.23 cm)
Maximum Span: 19¾ in (50.16 cm)
Function: Surface-to-Air
Weight: 149 lbs (67 kg)
Warhead: 20 lbs, High Explosive
Guidance: Radio Controlled
First Use Date: 1969
Producer: Euromissile
Users: France, Germany, Argentina, Brazil, Iraq, Nigeria, Qatar, Spain, Venezuela, Libya
Status: Tested, operational outside of the USA

Sequence of images showing the left ½ front view of an XM975 tracked vehicle firing an MIM-115 Roland surface-to-air missile.

Although intended for use by the US Army as it was assigned an MDS number, the MIM-115 was not picked up by the US military. Although Boeing did not win an American contract for the Roland, the original producers, Euromissile, a Franco-German combine, have made the MIM-115 for some nine nations. Launched from wheeled or tracked mobile launchers, the Roland engages both enemy aircraft and cruise missiles.

RIM-116 RAM (Rolling Airframe Missile)

Left ½ front view of an RIM-116 RAM (Rolling Airframe Missile) used for ship-to-air defense.

Left side view of an RIM-116 RAM (Rolling Airframe Missile) being fired from a shipboard mount.

Specifications

Length: 9 ft 4 in (284.48 cm)
Diameter: 5 in (12.7 cm)
Maximum Span: 1 ft 5.5 in (44.45 cm)
Function: Surface-to-Air Missile
Weight: 162 lbs (73 kg)
Warhead: 25 lbs, High Explosive
Guidance: Passive Radio Frequency / Infrared
First Use Date: 1993
Producer: Raytheon
Users: US Navy
Status: operational

Left ½ rear view of a missile mount firing an RIM-116 RAM (Rolling Airframe Missile). Note the rear of the missile tube opens when a missile is fired, with subsequent blast from the back of the mount.

The RIM-116 is a ship-to-air anti-cruise missile weapon also known as the Rolling Airframe Missile. The RAM is fired from the deck of surface ships from an 11- or 21-round launcher. The RAM has folding fins, which allow for closely spaced launch tubes. The RIM-116 can also be combined with the Phalanx Close-In Weapons System (CIWS), a 20mm Gatling gun system, as the SeaRAM Missile Defense System. Also a short range Anti-Ship Missile Defense (ASMD) weapon, the RIM-116 uses Sidewinder and Stinger components.

FQM-117 RCMAT/ARCMAT

Left side view of an FQM-117B ARCMAT target drone in flight.

Specifications

Length: 6 ft (182.88 cm)
Maximum Span: 5 ft 6 in (167.64 cm)
Function: Aerial Target
Weight: 8 lbs (3 kg)
Guidance: Radio Controlled
First Use Date: 1979
Producer: RS Systems
Users: US Army
Status: Operational

The FQM-117A RCMAT (Radio Controlled Miniature Aerial Target) was developed to provide soldiers with a low-cost aid for surface-to-air defensive weapons training. The RCMAT could be equipped with various devices to supplement the radar signature or to provide a heat source for infrared guided weapons. The FQM-117 was improved upon when it was upgraded to the ARCMAT (Augmented Radio Controlled Miniature Aerial Target) which added a three dimensional aspect to the device, allowing it to visually mimic actual aircraft. Capabilities were also enhanced.

The FQM-117B visually replicates a MiG-27 aircraft, but to a small scale (1/9). Powered by a model aircraft engine, this low-cost drone is intended to be the target of man-portable anti-aircraft weapons. The FQM-117C is a 1/9 scale representative of the F-16 Fighting Falcon, which is used to impress upon troops in training the importance of friend-or-foe recognition.

LGM-118 Peacekeeper

Right ½ front view of a full-scale mockup of the LGM-118 Peacekeeper (MX) missile.

View of the launch of an LGM-118 Peacekeeper (MX) missile. This was test FTM-18 launched March 19, 1989.

Specifications

Length: 70 ft 11 in (2161.54 cm)
Diameter: 7 ft 8 in (233.68 cm)
Function: Surface-to-Surface, Intercontinental Ballistic Missile
Weight: 194, 590 lbs (88,345 kg)
Warhead: (10) W87 Nuclear Warhead 300 KT
Guidance: Inertial
First Use Date: 1986
Producer: Lockheed Martin
Users: US Air Force
Other Designations: MX
Status: Was operational, withdrawn from use

The LGM-118 was developed as a successor to the Minuteman, but due to the Start II treaty, the missile was deactivated in the mid-1990s, even as the Minuteman continues in service. The Peacekeeper was developed under the "Missile Experimental" or MX program, and was to have utilized a 12-MIRV warhead, although treaty limits were made to restrict that to 10 warheads when fielded. The now-deactivated Peacekeeper missiles are used in a peaceful role as Orbital Sciences converts the boosters into Minotaur launch vehicles.

LGM-118 Peacekeeper missile just as it leaves a silo during a test launch.

AGM-119 Penguin

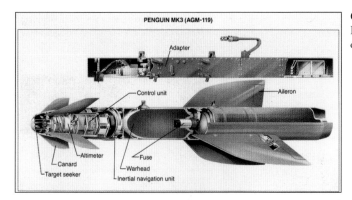

Cutaway view of the AGM-119 Penguin anti-ship missile, with component callouts.

Left ½ front view of four AGM-119 Penguin anti-ship missiles on loading dollies in front of a hardened shelter for a General Dynamics F-16A Fighting Falcon of the Royal Norwegian Air Force.

Left side view of a Sikorsky SH-60 Seahawk helicopter firing an AGM-119 Penguin anti-ship missile.

Specifications

Length: 10 ft 1 inch (307.34 cm)
Diameter: 11¼ in (28.57 cm)
Maximum Span: 3 ft 3 in (99.06 cm)
Function: Air-to-Surface
Weight: 820 lbs (372 kg)
Warhead: 250 lbs, High Explosive
Guidance: Inertial, with Infrared at terminal stage
First Use Date: 1993
Producer: Kongsberg Vaapenfabrikk, Norway
Users: Norway, US Navy, Turkey, Greece, Sweden
Status: Was operational, withdrawn from use in USA

The AGM-119 was until recently the only missile in use by the United States Navy produced by another country. First used in 1972 on Norwegian patrol boats, versions are now used on helicopters and fixed-wing aircraft. Inertial guided to a naval target, the Penguin then homes in on the target in the final stage of flight with infrared sensors.

AIM-120 AMRAAM

The first fully-guided prototype AIM-120A AMRAAM missile produced under the weapon's full-scale development program being inspected by Hughes Aircraft Company technician Steven P. Henchey prior to delivery from Hughes' Canoga Park, Calif., prototype laboratory. (The JAIM-120A designation indicates a test missile.) The advanced medium-range missile was delivered to Holloman Air Force Base, N.M., for a test launch over White Sands Missile Range, N.M. In background are other Hughes missiles, AIM-54 Phoenix (center) and AGM-65 Maverick.

FIRST GROUND LAUNCH – A Hughes Aircraft Company AIM-120 Advanced Medium-Range Air-to-Air Missile (AMRAAM) blasts off during the first ground test launch of the missile, conducted at the Point Mugu Missile Test Center in Southern California.

Two AIM-120 AMRAAM missiles are mounted on the outer left fuselage hard points on a McDonnell-Douglas F-15 Eagle fighter. In this location F-15s had previously mounted AIM-7 Sparrow missiles.

View of two Air Force weapons maintainers loading an AIM-120 onto a McDonnell-Douglas F-15 Eagle.

American Missiles

Right side view of a McDonnell-Douglas F-15 Eagle as it fires an AIM-120 Advanced Medium-Range Air-to-Air Missile (AMRAAM).

Left side view (sequence) of a General Dynamics F-16A Fighting Falcon as it fires an AIM-120 Advanced Medium-Range Air-to-Air Missile (AMRAAM).

Specifications

Length: 12 ft (365.76 cm)
Diameter: 7 in (17.77 cm)
Maximum Span: 2 ft 1 inch (63.5 cm)
Function: Air-to-Air
Weight: 335 lbs (152 kg)
Warhead: 50 lbs, High Explosive fragmentation
Guidance: Radar
First Use Date: 1991
Producer: Hughes, Raytheon
Users: US Air Force, US Navy, US Marine Corps, United Kingdom (plus over 30 others)
Status: Operational

Left rear view of a M1097 High Mobility Multi-Purpose Wheeled Vehicle (HMMWV) as it fires an Surfaced Launched Advanced Medium-Range Air-to-Air Missile (SLAMRAAM).

Jointly funded by the US Navy and US Air Force to serve as the medium-range missile to replace the Sparrow on Navy, Air Force and Marine Corps aircraft, the AIM-120 is known as the AMRAAM or Advanced Medium Range Air-to-Air Missile. The AIM-120 has an all-weather BVR (beyond visual range) capability, and is faster yet smaller and lighter than the AIM-7 Sparrow it replaces. The rocket motor of the AMRAAM is also made with a smaller smoke signature, making it harder to evade when fired at an opposing aircraft. The AIM-120 has been operationally deployed, and has seen success in downing enemy aircraft. A ground-launched variant known as SLAMRAAM has been fielded in some areas.

CQM/CGM-121 Seek Spinner

Right ½ front view of a CQM-121 drone sitting on a launch cradle/storage unit.

Right ½ front view of a CQM-121 drone during a rocket-assisted launch.

Specifications

Length: 7 ft (213.36 cm)
Height: 2 ft (60.96 cm)
Maximum Span: 8 ft (243.84 cm)
Function: Surface-to-Surface
Weight: 450 lbs (204 kg)
Warhead: High Explosive
Guidance: Programmed
First Use Date: 1981
Producer: Boeing
Users: US Air Force
Other Designations: Pave Tiger, Brave 200
Status: Tested

Right ½ rear view of a CQM-121 drone in flight.

The CQM-121 was a proposed anti-radar drone with a loiter capability. Once launched with rocket assist from the cradle arms of its storage container, the drone would fly a programmed path to the intended target area. Here the CQM-121 would orbit, loitering in the area until it detected radar emissions from enemy radar that might be switched off rather than left constantly on so as to escape detection. As most radar units rotate to acquire and track their targets, the name "Seek Spinner" was very appropriate. Once the CQM-121 had found a signal to track, it would home in and the direct hit with its 40-lb warhead would be enough to disable that particular radar unit. A missile battery may have more than one radar unit, but the storage container for the Seek Spinner would house several drones, so several could be programmed to hunt for radar in one area.

AGM-122 Sidearm

Right side view of the weapons pylon of a Bell AH-1 Cobra helicopter at the Naval Weapons Center, China Lake, CA, showing the AGM-122 Sidearm missile mounted.

"Awesome firepower displayed by McDonnell Douglas (Hughes) AH-64A Apache anti-armor helicopter includes: AIM-9 Sidewinder and AGM-122 Sidearm missiles (far left, far right) and Mistral air-to-air missiles (foreground); 1,200 rounds of 30mm ammunition; 72 2.75in/70mm rockets; 16 AGM-114 Hellfire anti-tank missiles (on inboard pylon stations); and wingtip mounted AIM-92 Stinger air-to-air missiles, along with external fuel tanks."

Specifications

Length: 9 ft 5 in (287.02 cm)
Diameter: 5 in (12.7 cm)
Maximum Span: 2 ft 1 inch (63.5 cm)
Function: Air-to-Surface
Weight: 195 lbs (88 kg)
Warhead: 21 lbs, High Explosive
Guidance: Radar Seeking
First Use Date: 1987
Producer: Motorola
Users: US Marine Corps
Status: Was operational, withdrawn from use
Family: Sidewinder

The AGM-122 combined an AIM-9C seeker modified for use in the anti-radiation role (and its fins) with the remainder of an AIM-9M Sidewinder. This reuse of a missile that has been replaced by a newer version can cause some identification confusion, as the upgraded casings called the Sidearm (Sidewinder Anti-Radiation Missile) looks externally much like the Sidewinder. The AGM-122 was developed by Naval Weapons Center, China Lake and produced by Motorola. The Sidearm was intended for use on Marine Corps AH-1W Cobra helicopters and the AV-8 Harrier.

AGM-123 Skipper II

Left ½ front view of the GBU-16 Paveway II Laser Guided Bomb. This weapon, modified with a rocket engine on the rear fin unit is redesignated the AGM-123 Skipper II.

US Air Forces weapon loaders load an AGM-123 Skipper air to surface missile onto a F-16C Fighting Falcon aircraft of the 555th Fighter Squadron prior to NATO air strikes on Serbian targets surrounding Sarajevo, Bosnia-Herzegovina.

Specifications

Length: 14 ft (426.72 cm)
Diameter: 1 ft 7½ in (49.53 cm)
Maximum Span: 5 ft 3 in (160.02 cm)
Function: Air-to-Surface
Weight: 1, 283 lbs (582 kg)
Warhead: 1,000 lbs, High Explosive
Guidance: Laser
First Use Date: 1985
Producer: Emerson Electric
Users: US Navy
Status: Was operational, withdrawn from use

Searching for a preceding namesake missile to the Skipper II will prove fruitless; the AGM-123 is so named in part as a salute to the laser guided bomb known as the Paveway II that the Skipper II is based upon.

The ever-fertile minds at the Naval Weapons Center at China Lake, California developed the AGM-123 to create a ready made laser-guided missile from components at hand. It combined the guidance and control set from the Paveway II, a 1,000-lb bomb and a rocket motor from a retired AGM-45 Shrike. The Paveway II flight controls are not smooth; they pulsed, which caused abrupt changes in the flight path. This was said to have caused the missile to "skip" in flight, much like a flat stone tossed across the surface of a pond. Skipper II missiles were retired after Operation Desert Storm. The rocket motors were scrapped, but other components were modified for use in laser-guided bombs.

AGM-124 Wasp

"SEEK AND STING – A technician at Hughes Aircraft Company's Missile Systems Group in Canoga Park, Calif., adjusts the wings on a full-scale model of the AGM-124 Wasp air-to-surface missile."

Specifications

Length: 5 ft (152.4 cm)
Diameter: 8 in (20.32 cm)
Maximum Span: 1 ft 8 in (50.8 cm)
Function: Air-to-Surface
Weight: 125 lbs (56 kg)
Warhead: High Explosive, Shaped Charge
Guidance: Programmed/Radar
First Use or Design Date: 1983
Producer: Hughes
Users: US Air Force
Status: Tested

The AGM-124 Wasp air-to-surface missile was developed for the US Air Force as an anti-armor missile that would have been launched from an aircraft weapons pod either singly or in a "swarm" of up to 10 or more. It was to have had a "lock-on after launch" capability, meaning the aircrew was not required to see and designate a target for the missile before it was fired. After flying a programmed route, the Wasp's radar seeker was to identify armored targets independent of the launching aircraft and would guide the weapon to an individual target. The autonomous targeting capability was to greatly increase the attacking aircraft's chances of survival, since it would not be exposed in a normal fashion to any enemy air defenses. The Wasp was not used operationally.

UUM/RUM-125 Sea Lance

Sea Lance (UUM-125), the anti-submarine warfare standoff weapon developed by Boeing Aerospace Company, will provide the US Navy's attack submarine fleet with a new missile for defense against hostile counterparts. Here, an engineering model is being loaded aboard a fleet attack submarine as part of a series of tests in Washington's Puget Sound. The model is a precisely weighted and sized copy of the actual missile and its launch capsule.

Specifications

Length: 20 ft (609.6 cm)
Diameter: 1 ft 9 in (53.34 cm)
Function: Surface-to-Ssurface (submarine to submarine)
Weight: 3,100 lbs (1,407 kg)
Warhead: Mk 50 Torpedo 100 lbs, High Explosive, Shaped Charge
Guidance: Digital-Inertial
First Use Date: 1985
Producer: Boeing
Users: US Navy
Status: Tested

The Sea Lance was a missile designed to provide the US Navy attack submarine fleet a defensive capability against other attack submarines. Launched "cold" from a forward torpedo tube, the UUM-125 would ignite its rocket motor once it cleared the surface. The Sea Lance was a standoff weapon, and much like air-launched standoffs, this allowed the launching submarine to remain outside the attack area of the opposing vessel. It was hoped that the UUM-125 could also be used with vertical launch by surface ships, but the Sea Lance was not adopted by the Navy.

An engineering model of Sea Lance (UUM-125), the Anti-submarine Warfare Standoff Weapon, is successfully launched by Boeing during a series of tests in Washington's Puget Sound. A full-scale replica of the launch capsule with a simulated missile inside, known as the "Iron Fish," was loaded aboard a US Navy fleet attack submarine for launching from one of its torpedo tubes. The launch proved that the capsule will rise to the surface in the correct attitude for launch of the missile. Sea Lance will provide the Navy's attack submarines with a replacement for older anti-submarine missiles in the late 1980s.

A Sea Lance (UUM-125) missile is launched from the shore of San Clemente Island, a Navy facility off San Diego, with a complete "all-up" missile, including avionics, flight software and all flight controls. The "dynamic test launch" was a significant milestone in the full-scale engineering development phase of Sea Lance by Boeing for the US Naval Sea Systems Command. The launch was made from a test fixture ashore rather than from a submarine. Sea Lance has been designated the common anti-submarine standoff weapon. It will be deployed on attack submarines and vertical launching system-equipped surface ships.

BQM-126

Left side view of a BQM-126 drone during launch.

Specifications

Length: 18 ft 6 in (563.88 cm)
Diameter: 1 ft 3 in (38.1 cm)
Maximum Span: 10 ft (304.8 cm)
Function: Aerial Target
Weight: 1,398 lbs (634 kg)
Guidance: Command Controlled, Programmed
First Use or Design Date: 1984
Producer: Beechcraft
Users: US Navy
Other Designations: Beech Model 997
Status: Tested

The BQM-126 was a proposed aerial target for use by the US Navy. It was to replace the BQM-34 Firebee and find use for air-to-air missile, surface-to-air missile, surface-to-air gunnery and anti-cruise missile training. After development and flight testing, a production contract was not awarded.

AQM-127 SLAT (Supersonic Low Altitude Target)

Right side view of an F-4 Phantom II with the AQM-127 SLAT on the centerline pylon.

Left side view of an F-4 Phantom II with an AQM-127 SLAT. Note the ramjet intake below the nose of the AQM-127.

Specifications

Length: 18 ft 2 in (553.72 cm)
Diameter: 1 ft 9 in (53.34 cm)
Function: Aerial Target
Weight: 2408 lbs (1,093 kg)
Guidance: Remote Control
First Use Date: 1987
Producer: Martin-Marietta
Users: US Navy
Status: Tested

Right side view of an F-4 Phantom II with an AQM-127 SLAT.

The AQM-127 was a drone intended to act as a supersonic target acting in the manner of a sea skimming anti-ship cruise missile. The AQM-127 used a combined rocket ramjet propulsion system to achieve speeds of Mach 2.5 while flying at an altitude of some 30 ft above sea level. This role was to cover tests and training for Aegis class anti-air naval vessels.

The SLAT was to be recoverable by parachute after each mission, with a projected life span of four missions. Tests were to be conducted using the F-4 Phantom II and DC-130 Hercules as launch platforms, with projected additional service use from the P-3 Orion, A-6 Intruder and F/A-18 Hornet aircraft.

The AQM-127 borrowed heavily on test and design data from the ASALM (Advanced Strategic Air Launched Missile) a missile program that was canceled before it received an MDS designation. Both the ASALM and SLAT used IRR (Integral Rocket/Ramjet) propulsion systems.

AQM-128

Specifications

None
Program cancelled

The AQM-128 was to have been an air-launched, non-recoverable, supersonic, subscale aerial target for the US Navy.

AGM-129 ACM

Left side view of an AGM-129 Advanced Cruise Missile (ACM) in flight.

Right side view, from below, of a Boeing B-52H Stratofortress, in flight, carrying six AGM-129 Advanced Cruise Missile (ACM) under each wing.

Specifications

Length: 20 ft 10 in (635 cm)
Diameter: 2 ft 5 in (73.66 cm)
Maximum Span: 10 ft 9 in (327.66 cm)
Function: Air-to-Surface
Weight: 3,500 lbs (1,589 kg)
Warhead: W80 Nuclear Warhead, 200 KT or High Explosive
Production date: 1990
Producer: General Dynamics/Raytheon
Users: US Air Force
Other Designations: Teal Dawn
Status: Was operational, withdrawn from use

The AGM-129 Advanced Cruise Missile was a subsonic cruise missile powered by a turbofan engine which was shielded to provide stealthy characteristics while in flight. The AGM-129 was fielded on B-52 Stratofortress bombers. Details on most aspects of the missile remain classified. The distinctive shovel-tip nose is the characteristic most easily seen when on the ground, but in flight the AGM-129 displays unique forward-swept composite wings. The AGM-129 was operated solely as a nuclear arm, and was withdrawn from use in 2008 under some controversy when live missiles were flown to retirement onboard a B-52. Production was halted in 1992 to meet disarmament figures after over 400 AGM-129s had been delivered.

AGM-130

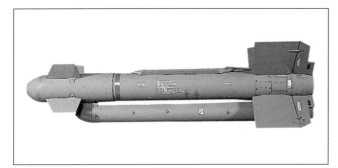

Side view of the AGM-130.

Front view of the AGM-130 showing the seeker head and the underslung propulsion unit.

Specifications

Length: 12 ft 10 1/2 in (392.43 cm)
Diameter: 1 ft 6 in (45.72 cm)
Maximum Span: 4 ft 11 in (149.86 cm)
Function: Air-to-Surface
Weight: 2,917 lbs (1,324 kg)
Warhead: 530 or 950 lbs, High Explosive, Shaped Charge
Guidance: Inertial/GPS, terminal television or imaging infrared seeker
First Use Date: 1994
Producer: Rockwell International, Boeing
Users: US Air Force
Status: Was operational, withdrawn

The AGM-130 was designed for the F-111F and F-15E aircraft. Derived from the GBU-15 guided glide bomb, the AGM-130 has navigation systems and propulsion, which make it a missile rather than a glide bomb. Depending on the warhead, the missile can either be the AGM-130A with a Mk-84 blast warhead, AGM-130B with a SUU-54 Cluster Bomb Dispenser, or a AGM-130C with a BLU-109 penetrator warhead. First used in January 1999 during Operation Northern Watch against Iraq's SAM sites. It was also used later in 1999 during Operation Allied Force against Serbia.

AGM-130 just after release from an F-15E aircraft.

AGM-131 SRAM II

Right ½ rear view as technicians on the Boeing Aerospace & Electronics AGM-131 Short Range Attack Missile (SRAM) II program work in the Systems Integration Laboratory, using the SRAM II Electronic System Test Set and a Missile Electronics System Simulator on a Ground Test Missile in preparation for a factory acceptance test.

Right ½ front view as a full-scale mockup of the Boeing AGM-131 Short Range Attack Missile (SRAM) II is positioned for checkout with the Electronic System Test Set.

Specifications

Length: 10 ft 5 in (317.5 cm)
Diameter: 1 ft 3¼ in (38.73 cm)
Function: Air-to-Surface
Weight: 2000 lbs (908 kg)
Warhead: W89 Nuclear Warhead, 200 KT
Guidance: Inertial
Design Date: 1980s
Producer: Boeing
Users: US Air Force
Other Designations: TASM, SRAM-T
Status: Built, not flight tested

The AGM-131 was a development to replace the AGM-69 SRAM or Short Range Attack Missile. This nuclear armed missile would be used as a standoff weapon which would allow the transporting bomber to release the missile at a short distance from the target, eliminating the need to over fly the target after release. This could be considered a bridge between a traditional gravity bomb and a cruise missile that would be launched at a distance from the target. Due to the collapse of the Soviet Union and the reductions in nuclear arms, the SRAM II was canceled, just as examples were being readied for flight tests. The missile would have been carried by the B-1 and B-2 bombers.

AIM-132 ASRAAM (Advanced Short Range Air-to-air Missile)

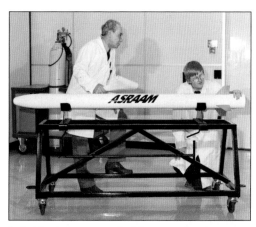

Left ½ front view of two AIM-132 Advanced Short Range Air-to-Air Missiles (ASRAAM) mounted on launch rails on the left wing of a General Dynamics F-16 Fighting Falcon.

Left side view as two technicians examine the AIM-132 Advanced Short Range Air-to-Air Missile (ASRAAM).

Specifications

Length: 9 ft (274.32 cm)
Diameter: 6½ in (16.5 cm)
Maximum Span: 1 ft 5¾ in (45.08 cm)
Function: Air-to-Air
Weight: 191 lbs (86 kg)
Warhead: 22 lbs, High Explosive, Fragmentation
Guidance: Infrared
First Use Date: 1998
Producer: Matra BAe Dynamics, Bodenseewerk Geraetetechnik
Users: Royal Air Force, Royal Australian Air Force
Status: Operational, but not by USA

Another missile to be given an MDS number as there was initial US involvement in its development is the AIM-132. The ASRAAM (Advanced Short Range Air-to-Air Missile) has been procured by the Royal Australian Air Force as the replacement for the AIM-9 Sidewinder. As a "within visual range" missile, the ASRAAM uses infrared guidance to find the target. AIM-132 was intended to be a stablemate of the AIM-120 AMRAAM and to replace the AIM-9 as the shorter range missile of the advanced pair. The multinational program broke down in the late 1980s due in part to a perceived incompatibility with the launch rails used by the AIM-9 that are standard on the prospective launch aircraft such as the F-16 and F-18. Testing and production of the AIM-132 continued under BAe, and the missile has been acquired by the air arms of the UK and Australia.

UGM-133 Trident II

Aerial port bow view of the nuclear-powered strategic missile submarine USS *Tennessee* (SSBN-734) underway as it nears its new home port of Naval Submarine Base, Kings Bay, GA. Trailing the *Tennessee* is a Navy large harbor tug (YTB). The *Tennessee* is armed with UGM-133 Trident II missiles. Dated January 15, 1989.

Differences in size is dramatized by this view of mockups of the UGM-133 Trident II (D-5) (left) and UGM-96 Trident I Fleet Ballistic Missiles at Lockheed Missiles and Space Co. Trident I was deployed operationally in 1979 aboard the US Navy's FBM submarines. Trident II is in full-scale engineering development and will start deployment in late 1989 aboard the *Ohio*-class of Trident submarines.

The first test-launched UGM-133 Trident II (D-5) Fleet Ballistic Missile roars from launch pad 46A, Cape Canaveral at 10:25 A.M. (EST) on Thursday, January 15, 1987. The flight missile, designated D5X-1 for "D5-experimental-first" met all parameters as it sped southeasterly down the Eastern Test Range to a planned impact area in the South Atlantic.

Specifications

Length: 44 ft 4 in (1351.28 cm)
Diameter: 6 ft 11 in (210.82 cm)
Function: Underwater-to-Surface, sea launched ballistic missile
Weight: 129,800 lbs (58,930 kg)
Warhead: (10-15) Nuclear Warhead, 300-475 KT
Guidance: inertial
First Use Date: 1987
Producer: Lockheed
Users: US Navy, Royal Navy
Other Designations: D-5
Status: Operational

The final development of the sea-launched ballistic missile is the Trident II, which was first flown in 1987, and first launched from a submarine in 1989. Initial trials discovered a flaw when the missile left the submarine, but a slight redesign and strengthening the area around the nozzle was a fix. The UGM-133 was designed to attack hardened targets, and was equipped with from ten to fifteen warheads of over 300 kilotons each. As part of the reduction in the numbers of nuclear weapons, President George H.W. Bush halted production of the Trident II in 1992, which later resumed. The UGM-133 is carried aboard American *Ohio*-class and the Royal Navy's *Vanguard* class strategic missile submarines.

Launch of UGM-133 (D-5) Trident II missile (test PEM-1) from the nuclear-powered strategic missile submarine USS *Tennessee* (SSBN-734) March 21, 1989 at 11:20 EST. This was not a successful launch.

Full-scale mockup of the Small ICBM, later designated the MGM-134 Midgetman missile.

Launch of Test Launch FIM-1, the firing of an MGM-134 Midgetman missile. The MGM-134 was the "Small ICBM" program, terminated shortly afterwards. The launch took place May 11, 1989.

MGM-134 SICBM

Specifications

Length: 46 ft (1402.08 cm)
Diameter: 3 ft 10 in (116.84 cm)
Function: Surface-to-Surface ICBM
Weight: 30,000 lbs (13,620 kg)
Warhead: (1) W87 Nuclear Warhead, 475 KT
Guidance: Inertial
First Use Date: 1989
Producer: Martin-Marietta
Users: US Air Force
Status: Tested

The MGM-134 was to have been a small three-stage ICBM that utilized the inertial guidance and control system from the Peacekeeper missile. Known unofficially as the "Midgetman" missile, one launcher proposal was via truck-based hardened mobile launcher. As fixed ICBM silos are themselves subject to nuclear attack, a land-based mobile ICBM was an attractive option as a nuclear deterrent. The program was cancelled in post-Cold War reductions, after tests flights had been made.

ASM-135 ASAT (Anti-Satellite Missile)

An ASM-135 Anti-Satellite Missile (ASAT), on a bomb loader, is fit-checked before being attached to the body of a McDonnell-Douglas F-15 Eagle aircraft.

Left side view of a McDonnell-Douglas F-15A Eagle in flight carrying an ASM-135 ASAT (Anti-Satellite) missile on the fuselage centerline.

Specifications

Length: 17 ft 9½ in (542.29 cm)
Diameter: 1 ft 8 in (50.8 cm)
Function: Air-to-air
Weight: 2,600 lbs (1,180 kg)
Warhead: Kinetic Warhead
Guidance: Radar
First Use or Design Date: 1985
Producer: Vought
Users: US Air Force
Status: Tested

Left side view of a McDonnell-Douglas F-15A Eagle in flight just as it has pitched up at a high angle of attack and released an ASM-135 ASAT (Anti-Satellite) missile from the fuselage centerline.

The controversial ASAT (or Anti-Satellite Missile) program was developed to counter a perceived threat by Soviet anti-satellite systems using "killer satellites." The program was dogged by those who claimed it simply would not work, and by those who felt it was just raising the ante in an ever-increasing arms race. But in 1979 Vought was given a contract to build an Anti-Satellite Missile, later designated ASM-135A.

Using components from existing systems married to new technology, the ASM-135 was carried by McDonnell-Douglas F-15 Eagles during a test program flown from Edwards AFB in California. On Sept 13, 1985, doubts about the system were put to rest when during a live-fire test the ASAT successfully destroyed an old communications satellite.

The ASM-135 used a Lockheed first stage motor from a SRAM (AGM-69) missile, and the second stage was adapted from the fourth stage of a Scout launch vehicle known as the Altair III. The warhead was a small self-propelled infrared Miniature Homing Vehicle or MHV. The MHV was flown close to the target by the ASM-135, and when released, guided itself to destruction by impact with the target vehicle. Some might claim that the ASM-135 is really a three-stage missile, using the F-15 as the first stage to altitude. Succumbing not to technology but to political pressures, the ASM-135 was cancelled in 1988, but after planning had begun for the acquisition of operational missiles and the assignment of F-15s to perform this specialized task.

AGM/BGM-136 Tacit Rainbow

Right side view of an AGM-136A Tacit Rainbow missile carried by a Grumman A-6E Intruder attack airplane in flight.

Right ½ rear view of an AGM-136A Tacit Rainbow missile in flight, banking away from the camera.

Specifications

Length: 8 ft 4 in (254 cm)
Diameter: 2 ft 3 in (68.58 cm)
Maximum Span: 5 ft 2 in (157.48 cm)
Function: Air-to-Surface, Surface-to-Surface anti-radiation
Weight: 430 lbs (195 kg)
Warhead: 40 lbs, High Explosive
Guidance: Programmed, radar seeker.
First Use Date: 1984
Producer: Northrop
Users: US Navy, US Air Force
Status: Tested

Tacit Rainbow was an unmanned aerial vehicle (UAV) designed to autonomously attack various enemy emitters, such as radar sites. The AGM-136 was to have been air or ground launched, and had the ability to loiter in a target zone until emissions were detected, at which point it would seek the target. If the targeted radar were switched off, it would resume loitering until a new target was acquired. The Tacit Rainbow vehicle would continue this cat-and-mouse until its fuel was expended or a target was destroyed. The AGM-136 program was cancelled after test flights had been made, but before serial production began.

AGM/MGM-137 TSSAM (Tri Service Standoff Attack Missile)

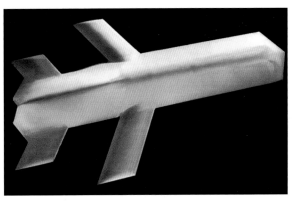

Right side view from above of a mockup of the AGM-137 TSSAM (Tri Service Standoff Attack Missile).

Specifications

Length: 14 ft (426.72 cm)
Maximum Span: 8 ft 4 in (254 cm)
Function: Air-to-Surface/Surface-to-Surface
Weight: 2,000 lbs (908 kg)
Warhead: High Explosive
Guidance: Inertial with GPS
Design Date: 1990s
Producer: Northrop
Users: United States
Status: Cancelled

Formulated on stealth technology to a larger extent than any other missile, the AGM-137 was to have served three services, hence the name as the TSSAM (Tri Service Standoff Attack Missile). The TSSAM was canceled in 1994, in part due to ballooning costs, in part due to the difficulties the services had in each separately funding the program on an equitable basis. The US Army (projected to use the ground-launched MGM-137) had dropped from the program earlier than the others (Marine Corps procuring through the US Navy) and the added expense of funding a larger share no doubt added to the pitfalls of funding. The missile had been flight tested before cancellation, and aspects of the missile would be used in the JASSM missile (AGM-158) program.

CEM-138 Pave Cricket

Specifications

None
Program cancelled

Proposed in the late 1980s, the CEM-138 was to have been a drone based on the same small airframe as the CQM-121. As the Pave Cricket vehicle, it was to have been used as an aerial jammer, carrying equipment to interfere and block radar signals from enemy transmitters on air defense systems. The program was cancelled.

RUM-139 VL-ASROC (Vertical Launch Anti-Submarine Rocket)

Naval Weapons Center China Lake, CA – A series of four views as a RUM-139 Vertical Launch ASROC (VL ASROC) missile is fired from its Vertical Launch System (VLS) during tests.

Naval Weapons Center China Lake, CA – A Vertical Launch System (VLS) magazine is prepared for testing of a RUM-139 Vertical Launch ASROC (VL ASROC) missile.

Specifications

Length: 16 ft ½ inch (488.95 cm)
Diameter: 1 ft 2 in (35.56 cm)
Maximum Span: 2 ft 3¼ in (69.215 cm)
Function: Surface-to-Surface (anti-submarine)
Weight: 1,408 lbs (639 kg)
Warhead: Mk 46 Torpedo, 98 lbs, High Explosive
Guidance: Inertial
First Use Date: 1993
Producer: Loral (Lockheed-Martin)
Users: US Navy, Japan
Status: Operational

A vast improvement over the RUR-5 ASROC, the RUM-139 had a more sophisticated delivery system for the torpedo payload the weapon carried for its anti-submarine role. The Vertical Launch ASROC (VL-ASROC) was no longer an unguided rocket, but a missile guided by a digital autopilot coupled with an inertial guidance package. Shortly after streaking from the missile container, the RUM-139 slew over in the direction of the target, and changes its angle yet again to stay low so that adverse winds do not affect the accuracy of the missile. At a calculated point, the torpedo falls away, to enter the water and move to the target.

A RUM-139 Vertical Launch ASROC (VL ASROC) missile is fired from its Vertical Launch System (VLS) during shipboard testing.

MGM-140 ATACMS (Army Tactical Missile System)

Right ¼ front view of an MGM-140 ATACMS (Army Tactical Missile System) missile in front of its transporter-launcher.

Right ¼ front view of the launch of an MGM-140 ATACMS (Army Tactical Missile System) missile from its transporter-launcher.

Specifications

Length: 13 ft (396.24 cm)
Diameter: 2 ft (60.96 cm)
Maximum Span: 4 ft 7 in (139.7 cm)
Function: Surface-to-Surface
Weight: 3681 lbs (1,671 kg)
Warhead: Bomblets
Guidance: Inertial
First Use Date: 1998
Producer: Lockheed
Users: US Army
Status: Operational

The MGM-140 is a single-stage ground-launched missile designed to intercept behind the lines forces attempting to reinforce troops and armor on the battlefield. The MGM-140 Army Tactical Missile System (ATACMS) missiles are fired from the MLRS (Multiple Rocket Launcher System) M270 and M270A1 weapons platform. To date it is the only long-range (more than 100 miles) surface-to-surface missile fired by the US Army in combat. Developed under a program known as Assault Breaker, the ATACMS is launched from the M270 carriage containing two MGM-140s, as opposed to twelve unguided MLRS rockets.

ADM-141 TALD

Left ½ front view of a McDonnell-Douglas F-18 Hornet fighter carrying a load of ADM-141A TALD decoys on underwing pylon racks.

Right ¾ rear view, from below, of a Lockheed S-3 Viking in flight carrying an ADM-141A TALD decoy on the right underwing pylon.

Right ¾ rear view of an ADM-141A TALD decoy in flight. Large number 13 painted on the side of the decoy.

Right side view of an ADM-141C ITALD decoy in flight. Composite studio image of the decoy superimposed on a sky background to give the appearance of flight.

Specifications

Length: 7 ft 8 in (233.68 cm)
Diameter: 1 ft 10 in (55.87 cm)
Maximum Span: 5 ft 11 in (180.34 cm)
Function: Aerial Decoy and Target
Weight: 200 to 400 lbs (91-181 kg)
Guidance: Programmed
First Use Date: 1991 (ADM-141B)
Producer: IAI, Brunswick
Users: US Navy, US Marine Corps, Israel
Status: Operational

The ADM-141 is an air-launched decoy, based on the Israeli Samson decoy, flown either unpowered as the ADM-141A/B TALD (Tactical Air Launched Decoy) or the ADM-141C ITALD (Improved Tactical Air Launched Decoy). The ADM-141 can either reflect radar when acting as a target, dispense chaff as a decoy, or emit infrared heat source to mimic aircraft of missiles. The TALD can be used as a training tool or to aid in the defeat of enemy air defense systems by simulating other weapons. ADM-141s were employed at the beginning of Operation Desert Storm to spoof Iraqi radars into turning on so that they could be struck by volleys of AGM-88 HARMs.

AGM-142 Raptor

Left ½ rear view, from below, of a Boeing B-52H Stratofortress carrying two AGM-142 Raptor missiles under the left wing.

Side view of an AGM-142 precision-guided missile on an F-16 Fighting Falcon, during testing at Eglin Air Force Base.

Specifications

Length: 15 ft 10 in (482.59 cm)
Diameter: 1 ft 9 in (53.34 cm)
Maximum Span: 5 ft 8 in (172.72)
Function: Air-to-Surface
Weight: 3,000 lbs (1,362 kg)
Warhead: 800 lbs, High Explosive
Guidance: Television and Infrared
First Use Date: 1985
Producer: Lockheed-Martin
Users: US Air Force, Australia, Israel, South Korea
Other Designations: Popeye , Have Nap
Status: Was operational, withdrawn from use in USA

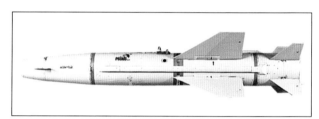

Side view of the AGM-142 Raptor missile.

Powered by a solid-fuel rocket motor and guided through electro-optical command, the Raptor missile is a stand-off air-to-surface missile normally employed against targets such as power stations and their associated structures, radar sites, petroleum processing plants, and other high-value targets that may have heavy anti-air defense systems. The stand-off capability allows the AGM-142 to strike such targets with the launch aircraft controlling for a distance, or it may hand off control to a second aircraft while it departs the target area. Originally developed from the Israeli "Popeye" under the name "Have Nap," the AGM-142 was carried by the B-52 and later tested (but not fielded) by the name "Have Light" on the F-16. During Operation Desert Storm the Raptor was not used as it was deemed politically insensitive to use an Israeli-developed weapon in that conflict.

MQM-143 RPVT

Specifications

Length: 10 ft (304.8 cm)
Height: 11 in (27.93 cm)
Maximum Span: 9 ft 10 in (299.71 cm)
Function: Aerial Target
Weight: 42 lbs (19 kg)
Guidance: Radio Control
First Use or Design Date:
Producer: Continental RPVs, Griffon Aerospace
Users: US Army

No image available

Much like the FQM-117 ARCMAT, the MQM-143 is a scale version of a larger aircraft to provide a target for troops training with anti-aircraft weapons, yet at the same time train them in aircraft recognition. In this case, the MQM-143 is a one-fifth scale version of the MiG-27 Flogger fighter. The US Army Missile Command is the contractor of record according to the DoD Designation list, although firms such as Continental RPVs and Griffon Aerospace have actually produced these target drones.

ADM-144

Specifications

None

Program cancelled.

BQM-145 Peregrine MR UAV (Medium Range Unmanned Aerial Vehicle)

A BQM-145 Peregrine is launched from a ground platform with a RATO unit.

Specifications

Length: 18 ft 4 in (558.8 cm)
Height: 33 ft 10 in (1031.24 cm)
Maximum Span: 10 ft 6 in (320.04 cm)
Function: Aerial Reconnaissance Drone
Weight: 2,000 lbs (908 kg)
Guidance: Autonomous or remote controlled
First Use Date: 1992
Producer: Teledyne Ryan, Northrop-Grumman
Users: US Air Force, US Navy, US Marine Corps
Other Designations:
Status: cancelled

Launched from the ground with RATO assist or from F-16 or F/A-18 aircraft, the BQM-145 was to perform as an unmanned aerial reconnaissance platform. Flying at subsonic speeds, the BQM-145 could fly a preprogrammed mission profile, or could have been flown by an operator by remote control. The onboard Global Positioning System (GPS) and Inertial Navigation System (INS) provided feedback in location and altitude to provide precise tracking and feedback. Upon completion of the mission the MRUAV would be recovered by parachute or parafoil at land or sea, or could be snatched mid-air by helicopter. The BQM-145 was to have carried the ATARS (Advanced Tactical Airborne Reconnaissance System) Electro-Optical and Forward Looking Infrared (FLIR) but when that program was cancelled, it was a vehicle in search of a purpose. Naval support ended, followed by a cessation of Air Force funding, effectively ending the program.

MIM-146 ADATS

Left ½ front view of an ADATS MIM-146 missile.

Left ½ front view of an ADATS MIM-146 missile turret equipped Bradley personnel carrier.

Specifications

Length: 6 ft 6 in (198.12 cm)
Diameter: 6 in (12.23 cm)
Maximum Span: 1 ft 7½ in (49.53 cm)
Function: Surface-to-Air/Surface-to-Surface
Weight: 112 lbs (51 kg)
Warhead: 27 lbs, High Explosive, Shaped Charge
Guidance: Laser
Design Date: 1980s
Producer: Oerlikon
Users: US Army
Status: Tested, operational outside of the USA

A laser-guided anti-aircraft missile, the ADATS is also designed for use in the anti-tank role. The MIM-146 is used against low-flying targets where use of radar guidance would be problematical. When enemy helicopters and low flying fixed wing aircraft use terrain masking to hide from defenses, the laser guidance can precisely target the MIM-146. The ADATS or Air Defense Anti-tank System uses laser beam relay for missile control. Although relatively small and quick (speed close to Mach 3) the MIM-146 did have difficulties in some weather conditions. The US Army declined to procure the ADATS, however Canada and Thailand are users of the missile. A navalized variant called Sea Sprint has been offered.

Left ½ front view of an ADATS MIM-146 missile as it leaves the missile tube during a test firing. The ADATS launcher is mounted on a modified M113 personnel carrier.

BQM-147 Dragon

BQM-147 Dragon drone is launched during testing, the large delta wing clearly visible as it leaves the launch trailer.

Specifications

Length: 5 ft 3½ in (161.29 cm)
Height: 1 ft 7 in (48.26 cm)
Maximum Span: 8 ft 2½ in (250.19 cm)
Function: Mini-RPV (Remotely Piloted Vehicle) drone
Weight: 91 lbs (41 kg)
Guidance: Radio Control
First Use Date: 1988
Producer: BAI Aerosystems
Users: US Army, US Marine Corps
Other Designations: Exdrone (Expendable Drone)
Status: Was operational, withdrawn from use

Rail-launched, the BQM-147 looks like a large bat with its distinctive delta wing. But this aerial creature can perform a large variety of missions including reconnaissance, signals relay, even payload delivery if the payload can fit in the space that is normally used by mission-specific electronics such as cameras. The BQM-147 was first used operationally during the 1991 Gulf War.

FGM-148 Javelin

Left ½ front view of an Army soldier manning an AAWS-M FGM-148 missile launcher.

Left ½ front view of an AAWS-M FGM-148 missile. – TEST MISSILE – Hughes Aircraft Company technician G.R. Patton checks over one of the guided test vehicles to be launched during the technology demonstration phase of the US Army's Advanced Anti-tank Weapons System – Medium (AAWS-M) program. The tests are being conducted at the Army's Redstone Arsenal in Huntsville, Ala. Hughes' concept is an imaging infrared-guided lightweight missile featuring a fiber-optic data link that transmits two-way information between the missile and launcher. The wings and control surfaces fold into the missile before launch. The missile can be locked onto a target either before launch, in a fire-and-forget mode, or after launch. The operator can stay hidden and observe what the missile's focal plane array seeker "sees," enabling him to select a target after firing, or to update the aim point on a selected target.

Left side view of Army soldiers firing an AAWS-M FGM-148 missile launcher, the missile has just barely left the launch tube.

Specifications

Length: 3 ft 6 in (106.67 cm)
Diameter: 5 in (12.7 cm)
Maximum Span: 1 ft 4 in (40.64 cm)
Function: Surface-to-Surface anti tank
Weight: 26 lbs (12 kg)
Warhead: 19 lbs, Tandem High Explosive, Shaped Charge

Guidance: Imaging Infrared
First Use Date: 1996
Producer: Lockheed/Raytheon
Users: US Army, US Marine Corps
Other Designations: AAWS-M (Advanced Anti-tank Weapons System – Medium)
Status: operational

Developed as joint venture between Texas Instruments and Martin Marietta (now Raytheon and Lockheed Martin respectively) the Javelin replaced the Dragon in the man-portable anti-tank role. Although the FGM-148 can be used in direct fire mode against armor, fortifications like bunkers or even low-flying aircraft like helicopters, the Javelin can be directed into an arcing flight profile that will allow it to strike armored vehicles from above, where the armor is normally thinnest. The guidance suite coupled with an imaging infrared seeker allows for fire-and-forget, which enables the operator to hide from enemy retaliation after firing the FGM-148.

PQM-149 UAV-SR Sky Owl

Specifications

Length: 13 ft 6 in (411.48 cm)
Maximum Span: 24 ft (731.52 cm)
Function: UAV (unmanned aerial vehicle) - Drone
Weight: 1,250 lbs (567 kg)
Guidance: Radio Controlled
First Use Date: 1986
Producer: McDonnell-Douglas
Users: US Army
Status: Cancelled

An unsuccessful entrant in a UAV (Unmanned Aerial Vehicle) competition ultimately won by the BQM-155, the PQM-149 was offered by McDonnell-Douglas, based on the R4E-50 Sky Eye from BAE Systems.

PQM-150

Specifications: None

Project cancelled: No known images

Designation approved in September of 1990 for a short-range reconnaissance, surveillance and target acquisition drone to be developed by the US Army together with the US Navy.

FQM-151 Pointer

Two US Marines carry an FQM-151 Pointer drone and its controller to a launch position.

Right ½ rear view as a US Marine readies a FQM-151 Pointer drone for launch.

Specifications

Length: 6 ft (182.88 cm)
Maximum Span: 9 ft (274.32 cm)
Function: Mini-UAV (Unmanned Aerial Vehicle) – battlefield surveillance
Weight: 9.2 lbs (4 kg)
Guidance: Radio Controlled
First Use Date: 1988
Producer: Aerovironment
Users: US Army, US Marine Corps
Status: Was operational, withdrawn from use

A mini-UAV, the FQM-151 resembles the radio-controlled airplanes of hobbyists, but the Pointer is flown over a hostile battlefield, not a recreational area. The PQM-151 uses an electric motor, and the high efficiency batteries also supply power to the payload, which can be a B&W or color television camera or a night-vision television camera. The Pointer is hand-launched, and the small size allows for a pancake full stall landing in a small area after the motor is cut.

Left ½ front view of an FQM-151 Pointer drone in flight.

AIM-152 AAAM

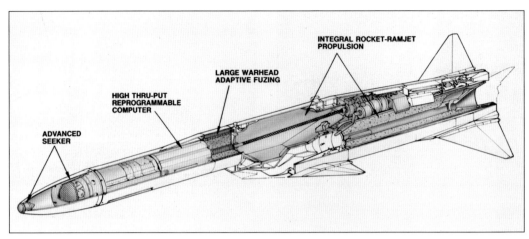

Left ½ front cutaway diagram view of an AAAM AIM-152 air-to-air missile.

Specifications

Length: 12 ft (365.76 cm)
Function: Air-to-Air
Warhead: 30-50 lbs, High Explosive, fragmentation
Guidance: inertial with radar
Design Date: 1990s
Producer: Hughes/Raytheon, GD/Westinghouse
Users: US Navy
Status: Canceled

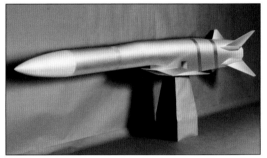

Left ½ front view of the Hughes/Raytheon entry for the AAAM AIM-152 air-to-air missile.

The AIM-152 AAAM (Advanced Air-to-air Missile) was a proposed long-range air-to-air missile to replace the AIM-54 Phoenix as used by the US Navy's F-14 Tomcat. Due in part to a significant decrease in the threat of attack by Soviet bombers, the AAAM program was canceled in 1992 before production began. No AIM-152s were built by either Hughes/Raytheon or General Dynamics/Westinghouse, the two teams that were competing for the project.

AGM-153

Specifications

None

US Air Force program cancelled before examples were built.

AGM-154 JSOW

Right side view from below of a McDonnell-Douglas F-18 Hornet fighter carrying two AGM-154 JSOW (Joint Stand Off Weapons) on the outboard wing pylons. Although designated as missiles, by the AGM designation, these are glide weapons.

View from below as an AGM-154 JSOW (Joint Stand Off Weapon) is about to penetrate a test target.

Specifications

Length: 13 ft 4 in (406.40 cm)
Diameter: 1 ft 1 in [square-sided] (33.02 cm)
Maximum Span: 8 ft 10 in (269.24 cm)
Function: Air-to-Surface
Weight: 1,065-1,500 lbs (483-681 kg)
Warhead: Bomblets or 500 lbs, High Explosive
Guidance: GPS/Inertial
First Use or Design Date: 1999
Producer: Raytheon
Users: US Navy, US Air Force
Status: Operational

The AGM-154 Joint Stand Off Weapon, or JSOW, is a modular weapon that can carry various types of warheads or weapons containers, can be powered or unpowered and can be guided by several means. All of these traits can be tailored to a specific mission, even to differences between missiles on the same delivery aircraft. The JSOW was designed to be carried by Air Force and Navy aircraft ranging from the F/A-18 to the B-2. However, the USAF pulled out of the program in 2005. The AGM-154C has infrared guidance at the terminal stage.

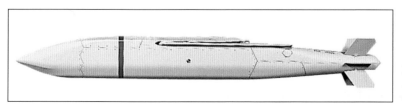

Left side view of the AGM-154 JSOW with wings folded.

BQM-155 Hunter

Left ½ front view of a BQM-155 Hunter drone during launch from a handling-launch trailer using rocket-assisted takeoff.

Front ¼ left view of a BQM-155 Hunter drone in flight.

Specifications

Length: 22 ft 9½ in (694.69 cm)
Height: 5 ft 5 in (165.1 cm)
Maximum Span: 29 ft 2½ in (890.27 cm)
Function: Drone – battlefield reconnaissance and target spotting
Weight: 1,600 lbs (726 kg)
Guidance: Programmed or Radio Controlled
Design Date: 1980s
Producer: TRW – IAI
Users: US Army, US Navy, US Marine Corps, France, Israel
Status: Operational

A joint venture between TRW and IAI (Israeli Aircraft Industries), the BQM-155 was developed to meet the needs for a short-range unmanned aerial vehicle (UAV-SR). The Hunter competed with the PQM-149 for the contract and ultimately won. The BQM-155 Hunter was later redesignated the RQ-5A in 1999.

RIM-156 Standard SM-2ER Block IV

View of a RIM-156 Standard SM-2ER Block IV missile as it leaves its vertical launcher.

Specifications

Length: 21 ft 6 in (655.32 cm)
Diameter: 1 ft 9 in (53.34 cm)
Maximum Span: 3 ft 6 in (106.67 cm)
Function: Surface-to-Air
Weight: 3,225 lbs (1,464 kg)
Warhead: High Explosive, fragmentation
Guidance: Semi-Active Radar
First Use Date: 1998
Producer: Raytheon
Users: US Navy
Status: Operational

The Standard Missile in yet another revised form is in US Navy service as the RIM-156. This extended-range version of the SM-2 uses a new booster to allow the use of the weapon from the Vertical Launch System (VLS), negating the requirement of the rail launchers of the "Terrier" ships. The new booster has no fins, and uses thrust vectoring for control of the missile while the solid rocket motor is burning during the boost phase.

MGM-157 EFOGM

Right ¼ front view of a modified Army weapons carrier that holds an eight-tube launcher for the MGM-157 EFOGM (Enhanced Fiber Optic Guided Missile).

MGM-157 EFOGM displayed as a cutaway showing internal details.

Specifications

Length: 6 ft 5 in (195.57 cm)
Diameter: 6½ in (16.5 cm)
Maximum Span: 3 ft 8¾ in (113.66 cm)
Function: Surface-to-Surface Anti-tank
Weight: 117 lbs (53 kg)
Warhead: High Explosive, Shaped Charge
Guidance: Optical fiber wire guided
First Use Date: 1997
Producer: Raytheon
Users: US Army
Other Designations: NLOS-CA, FOGM/NLOS-FAAD
Status: Tested

The Enhanced Fiber Optic Guided Missile (EFOGM) was a surface-to-surface guided missile for use primarily against armor and other mobile ground targets. Guidance was via an optical seeker head sending an image via a fiber-optic line to the operator. The MGM-157 missile could also be used to defend against low-flying aircraft such as helicopters. The EFOGM was over 6 ft long and

View of a test firing of a EFOGM MGM-157 missile.

used rectangular cruciform wings mounted close to the mid-body, with smaller fins at the rear. The missile was optimized for striking the target from the top, where armor is thinnest on most armored vehicles. The EFOGM was launched from a HMMWV (High Mobility Multi-Purpose Wheeled Vehicle) aka Hummer, the standard Army wheeled vehicle. The MGM-157 was not procured beyond the development phase.

AGM-158 JASSM (Joint Air-to-Surface Standoff Missile)

AGM-158 is handled by weapons technicians before loading onto a B-1B Lancer at Dyess Air Force Base, Texas. The Joint Air-to-Surface Standoff Missile (JASSM) is an air-to-surface, single warhead self-propelled missile.

Side view of an AGM-158 on a test flight while being paced by an observing General Dynamics F-16B Fighting Falcon.

Specifications

Length: 14 ft (426.72 cm)
Maximum Span: 7 ft 11 in (241.3 cm)
Function: Air-to-Surface
Weight: 2250 lbs (1,021 kg)
Warhead: 1,000 lbs, High Explosive
Guidance: GPS/Inertial
First Use Date: 1999
Producer: Lockheed Martin
Users: US Air Force, Royal Australian Air Force
Status: Operational

The AGM-158 low observable cruise missile known as the Joint Air-to-Surface Standoff Missile (JASSM) is an autonomous, long-range, conventional, air-to-ground, precision standoff missile for the US Air Force. The low-observable (stealth) aspects of the AGM-158 enhance its significant standoff range and helps keep aircrews well out of danger from hostile air defense systems. The JASSM is designed to destroy high-value, well-defended, fixed and re-locatable targets. In testing, the missile's mission effectiveness approaches single-missile target kill capability. The AGM-158B is under development, which is an Extended Range version, the JASSM-ER. The JASSM was originally a joint service program, but the US Navy withdrew to develop the AGM-84H/K, the SLAM-ER.

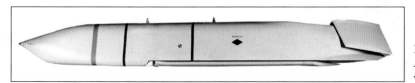

Left side view of an AGM-158 JASSM with wings folded.

AGM-159 JASSM (Joint Air-to-Surface Standoff Missile)

Image of the AGM-159 in flight.

Specifications

Dimensions: Unknown
Function: Air-to-Surface
First Use or Design Date:
Producer: McDonnell-Douglas
Users: US Air Force, US Navy
Status: Cancelled

Missile not produced, competition for the JASSM was won by the rival AGM-158. Dimensions of the AGM-159 were close to those of the AGM-158.

View of two AGM-159 JASSM missiles under the wing of an F-16 Fighting Falcon.

ADM-160 MALD (Miniature Air Launched Decoy)

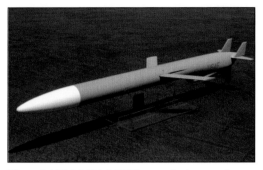
View of ADM-160A MALD on a display stand.

Later ADM-160B MALD under the wing of an F-16.

Specifications

Length: 7 ft 9¾ in (238.12 cm)
Maximum Span: 2 ft 1½ in (64.77 cm)
Function: Decoy to mimic manned aircraft
Weight: 100 lbs (45 kg)
Guidance: Pre-programmed
Producer: Northrop Grumman, Teledyne Ryan, Raytheon
Users: US Air Force
Status: Operational

The ADM-160 is a decoy designed to help US and allied air forces overwhelm enemy defenses. Developed by DARPA, the Miniature Air Launched Decoy (MALD) stimulates enemy air defenses by mimicking manned aircraft, confuses enemy air defenses by deceiving the true intent of the force, and saturating the enemy defenses by numerical advantage when added to an existing strike force. The ADM-160A was initially produced by Teledyne Ryan, which became part of Northrop-Grumman. Later variants were built by Raytheon, which appear slightly different in fin and wing configuration.

Early revision of the ADM-160B MALD at an exhibition.

RIM-161 Standard SM-3

The Arleigh Burke-class guided missile destroyer USS Hopper (DDG 70), equipped with the Aegis integrated weapons system, launches a RIM-161 Standard Missile (SM) 3 Block IA during exercise Stellar Avenger while under way in the Pacific Ocean July 30, 2009. The missile sucessfully intercepted a sub-scale, short-range, ballistic missile that was launched rom the Kauai Test Facility, Pacific Missile Range Facility at Barking Sands in Kauai, Hawaii. The exercise was the 19th successful intercept of 23 at-sea firings using the Aegis Ballistic Missile Defense program, including the destruction of a malfunctioning satellite above the earth's atmosphere in February 2008.

Specifications

Length: 21 ft 6 in (655.32 cm)
Diameter: 1 ft 1½ in (34.29 cm)
Maximum Span: 3 ft 6 in (106.67 cm)
Function: Surface-to-Air
Weight: Classified
Warhead: High Explosive, fragmentation
Guidance: Radar/Infrared
First Use Date: 1990s
Producer: Raytheon
Users: US Navy
Status: Operational

The newest version of the Standard Missile is the RIM-161. When carried on AEGIS-class destroyers and cruisers, the RIM-161 will provide the fleet with a ballistic missile defense against short to intermediate range ballistic missiles. The RIM-161 is a three stage missile, with the third stage taking a hit-to-kill infrared guided warhead to the target. The first and second stages are the same as the RIM-156 Standard Missile SM-2ER Block IV used for fleet area defense for airborne targets.

An RIM-161 Standard Missile Three (SM-3) is launched from the Mark 41 Vertical Launch System (VLS) onboard the US Navy (USN) TICONDEROGA CLASS: GUIDED MISSILE CRUISER (Aegis) USS LAKE ERIE (CG 70) at the Pacific Missile Range Facility, Kauai, Hawaii. The US Navy (USN) and The Missile Defense Agency (MDA) announced a successful flight test in the continuing development of the Sea-Based Midcourse (SMD) element of the Ballistic Missile Defense System.

RIM-162 ESSM (Evolved Sea Sparrow Missile)

Left ½ front view of an RIM-162 ESSM (Evolved Sea Sparrow Missile) as technicians perform final checkouts before delivery.

View of an RIM-162 ESSM as it leaves its vertical launcher.

Specifications

Length: 12 ft 1 inch (368.3 cm)
Diameter: 10 in (25.4 cm)
Function: Surface-to-Air
Weight: 615 lbs (279 kg)
Warhead: 66 lbs, High Explosive, fragmentation
Guidance: Radar with Inertial
First Use Date: 1997
Producer: Raytheon
Users: US Navy, Australia, Canada, Denmark, Germany, Netherlands and Norway
Other Designations: RIM-7PTC
Status: Operational

Initially developed as such a close follow-on to the Sea Sparrow that it was designated the RIM-7PTC, the Evolved Sea Sparrow Missile (ESSM) was later designated the RIM-162. With about twice the range of the RIM-7, the ESSM is compatible with the AEGIS shipboard weapons system, and the folding fins of the missile allow it to be contained in a four-tube shipboard launcher. The first production ESSM was delivered in September of 2002 to the US Navy, and they are currently deployed on *Arleigh Burke* class destroyers, and are being integrated into other vessels and new ships that enter service.

GQM-163 Coyote

Right ½ rear view of a GQM-163 Coyote missile just as it has cleared the launcher rail.

Right side view of a GQM-163 Coyote missile just after it has left the launcher.

Specifications

Length: 18 ft (548.64 cm)
Diameter: 1 ft 2 in (35.56 cm)
Function: Supersonic Target
First Use Date: 2003
Producer: Orbital Sciences
Users: US Navy
Status: Under testing

The GQM-163 Coyote is a Supersonic Sea Skimming Target (SSST) missile made to mimic the threat posed by an enemy anti-shipping cruise missile against US naval forces.

The Coyote is launched using a Mark 12 booster as used with the Standard Missile program. The missile proper uses solid fuel four-duct ramjet technology as the propulsion base, allowing speeds up to Mach 2.5.

MGM-164 ATACMS II (Army Tactical Missile System)

The MLRS (Multiple Rocket Launcher System) M270 carrier transported and launched the MGM-164 ATACMS II missile. Here the launcher is lowered while the M270 is underway."

View of BAT (Brilliant Anti-armor Technology) submunition with all the fins fully extended as normally seen after release from the MGM-164 warhead.

Specifications

Length: 13 ft (396.24 cm)
Diameter: 2 ft (60.96 cm)
Maximum Span: 4 ft 7 in (139.7 cm)
Function: Surface-to-Surface
Weight: 3,681 lbs (1,671 kg)
Warhead: BAT submunition
Guidance: Inertial
First Use Date: 2000
Producer: Lockheed
Users: US Army
Status: Cancelled

The MGM-164 is a single-stage ground-launched missile designed to defeat armored vehicles by use of the BAT submunition. The MGM-164 is known as the ATACMS (Army Tactical Missile System) Block II. Externally identical to the MGM-140 missile, was also fired from the MLRS (Multiple Rocket Launcher System) M270 and M270A1 weapons platform. The MGM-164 program has been cancelled.

The MGM-164 is the MGM-140 upgraded to Block II standard and carries the BAT (Brilliant Anti-tank) submunition. Like the earlier Army Tactical Missile System (ATACMS), it is fired from the MLRS (Multiple Rocket Launcher System) M270 and M270A1 weapons platform.

RGM-165 LASM

View of an RGM-165 LASM (Land Attack Standard Missile) as it rapidly approaches its impact point during a test.

Specifications

Length: 15 ft 6 in (472.44 cm)
Diameter: 1 ft 1½ in (34.29 cm)
Maximum Span: 3 ft (91.44 cm)
Function: Surface-to-Surface
Weight: 1,558 lbs (707 kg)
Warhead: High Explosive, fragmentation
Guidance: Inertial with GPS
First Use Date: 1997
Producer: Raytheon
Users: US Navy
Status: Operational

The RGM-165 is a Standard Missile previous configured for targeting airborne threats, and rebuilt with a inertial guidance suite that is aided by GPS (Global Positioning System). The new targets of the rebuilt missiles are land targets, to allow the weapon to fill the role of the intermediate weapon in a three-layer approach to land attack. The cruise missile would fill the need for deep strike, the LASM the middle layer, and guided munitions fired from 5-in guns would take the short or shoreline layer. The ease at the integration of this weapon into the fleet is that as a reconfigured Standard Missile, the ships are already equipped with the proper launchers to fire the RGM-165.

MGM-166 LOSAT

Left side view of an MGM-166 LOSAT missile in flight. The Line-Of-Sight Anti-Tank Missile (LOSAT) was known as the Kinetic Energy Missile (KEM).

Front view of an MGM-166 LOSAT (Line Of Sight Anti-tank) missile in flight after leaving the launcher.

Specifications

Length: 9 ft (274.32 cm)
Diameter: 6 in (15.23 cm)
Maximum Span: 10 in (25.40 cm)
Function: Surface-to-Surface Anti-tank
Weight: 175 lbs (79 kg)
Warhead: High Density Penetrator
Guidance: Inertial
First Use Date: 1990
Producer: Lockheed-Martin
Users: US Army
Other Designations: KEM
Status: Cancelled

The LOSAT or Line of Sight Anti-tank missile was intended to replace the BGM-71 TOW missile in the US Army's anti-tank arsenal. LOSAT, designated the MGM-166, was a laser-guided weapon, in that the missile rode a laser beam to the target, rather than tracked a target that was illuminated by a laser. Targeting in this manner was a clue to LOSAT being a hypervelocity projectile, travelling at speeds of over 1 mile per second in flight. This manner of laser beam riding did cause problems in development, as the missile had difficulty tracking the laser beam through the gases of its own rocket exhaust plume.

BQM-167 Skeeter

Launch of a BQM-167.

The crew from *Florida Offshore* steadies and lowers the BQM-167 sub-scale drone to the cradle after a recovery demonstration July 22, 2009 in the waters off Tyndall Air Force Base, Fla. The ship used for recovery is one of only three 120-ft boats owned by the Air Force. The *Florida Offshore* crew is contracted through the 82nd Aerial Targets Squadron to help recover sub-scale drones after they are shot down during live-fire exercises.

Specifications

Length: 20 ft (609.60 cm)
Diameter: 2 ft (60.96 cm)
Maximum Span: 11 ft (335.28 cm)
Function: Aerial Target
Weight: 690 lbs (313 kg) empty
Guidance: Command Guided
First Use Date: 2001
Producer: Composite Engineering
Users: US Air Force
Status: Operational

The BQM-167 is a ground-launched high-speed aerial target. Known as the Skeeter, this drone is scheduled to replace other drones currently in service. A slender airframe with an underslung turbojet engine, the BQM-167 looks similar to other aerial targets, but is made from composite material rather than metals. Radar reflectors are built into the airframe to allow the radar signature of the Skeeter to be tailored to mimic certain vehicles. A new variant of the Skeeter has been selected by the US Navy to replace the BQM-74E as a high speed aerial target.

Right front view of a BQM-167 Skeeter target missile in flight.

MGM-168 ATACMS Block IVA (Army Tactical Missile System)

Test launch of the MGM-168 ATACMS Block IVA missile from the White Sands Proving Ground, New Mexico.

Specifications

Length: 13 ft (396.24 cm)
Diameter: 2 ft (60.96 cm)
Maximum Span: 4 ft 7 in (139.70 cm)
Function: Surface-to-Surface
Weight: Unknown
Warhead: 500 lbs, High Explosive
Guidance: Inertial
First Use Date: 2001
Producer: Lockheed
Users: US Army
Status: Operational

The MGM-168 is another weapon from the ATACMS (Army Tactical Missile System) family, this being the Block IV, which uses a single 500-lb warhead in place of the tennis ball-sized M74 bomblets of the MGM-140. Again as it is externally identical to the MGM-140 missile, it will be fired from the MLRS (Multiple Rocket Launcher System) M270 and M270A1 weapons platform.

AGM-169 JCM (Joint Common Missile)

Left ½ front view, from slightly below, during the test firing of an AGM-169 Joint Common Missile (JCM).

Front view of a Hughes AH-64 Apache attack helicopter on the ground in a hangar with two AGM-169 Joint Common Missiles (JCM) mounted on each of the outboard weapons pylons.

Specifications

Length: 5 ft 9⅞ in (175.33 cm)
Diameter: 7 in (17.77 cm)
Maximum Span: 1 ft ¾ inch (32.38 cm)
Function: Air-to-Surface
Weight: 108 lbs (49 kg)
Warhead: High Explosive, Shaped Charge or fragmentation
Guidance: Semi Active Laser, imaging infrared, millimeter wave radar
First Use Date: 2003
Producer: Lockheed-Martin
Users: US Army, Navy, Marine Corps
Status: Tested, cancelled

Designed to replace both the AGM-65 Maverick and AGM-114 Hellfire missiles, the Joint Common Missile (JCM) was to be used in the air to ground anti-armor role. The AGM-169 could be equipped with a warhead optimized for penetrating armor, or a warhead used against a "softer" target. In order to fit existing launch rails on Army, Navy and Marine Corps aircraft, the JCM is externally very much like the Hellfire it would replace, but has a seeker head much more capable in working in all weather and varying combat conditions. Initially canceled in 2004, the AGM-169 had received funding to maintain program viability, but is now considered dead as the JAGM, or Joint Air to Ground Missile is under development. The JAGM is slated to replace the two missiles theoretically replaced by the JCM, plus the BGM-71 TOW.

MQM-170 Outlaw

Rear view of the MQM-170 Outlaw showing the rear-mounted engine and pusher propeller.

MQM-170 Outlaw as seen on transport trailer/launcher.

Specifications

Length: 9 ft (274.32 cm)
Maximum Span: 13 ft 6 in (411.48 cm)
Weight: 130 lbs (59 kg)
Function: Aerial Target
Guidance: Radio Control
Producer: Griffon Aerospace
Users: US Army
Status: Operational

The MQM-170 Outlaw is an unmanned aerial vehicle used as recoverable target as the MQM-170A and as a special purpose expendable platform as the MQM-170B. The Outlaw is intended to provide the US Army needs for an aerial target for both gunnery and anti-air missile crews in training.

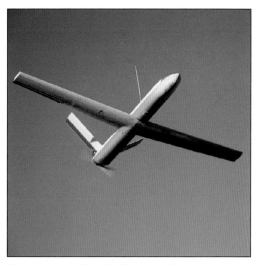

View of the MQM-170 Outlaw in flight.

MQM-171 BroadSword

Left front MQM-171 BroadSword on transport trailer/launcher.

Let rear view of MQM-171 BroadSword on transport trailer/launcher with launch rails extended. View of rear shows propulsion system and pusher propeller.

Specifications

Length: 13.78 ft
Diameter: 1.48 ft–1.67 ft
Maximum Span: 17.06 ft
Function: Aerial Target
Weight: 550 lbs (249 kg)
Guidance: Radio Control
First Use or Design Date: 2007
Producer: Griffon Aerospace
Users: US Army
Status: Operational

The MQM-171 BroadSword is an unmanned aerial vehicle used as a target to mimic the characteristics of tactical unmanned aerial vehicles that may be used against US forces. As a target, the BroadSword would be used to test new weapons systems.

Although similar in layout to the MQM-170 Outlaw, also produced by Griffon Aerospace, the MQM-171 is a larger vehicle. It is undergoing testing and has not yet been deployed to units.

FGM-172 Predator

Left ½ front view, from slightly below, during the test firing of an FGM-172 Predator. The missile has just left the launch tube.

Left side view of a Marine aiming an FGM-172 Predator missile from behind a low wall at a target. The shoulder fired weapon is man portable.

Specifications

Length: 35 in
Diameter:
Maximum Span:
Function: Surface-to-Surface, Anti-tank or assault
Weight: 21.3 lbs (9 kg)
Guidance: Inertial
First Use Date: 2001
Producer: Lockheed-Martin
Users: US Marine Corps
Other Designations: SRAW (Short Range Assault Weapon)
Status: Operational

The Predator is a shoulder-fired missile, originally used in the "A" variant by the US Marine Corps as an anti-armor weapon. Once fired, the missile would overfly the intended target and direct a shaped charge downward through the thinner armor normally on the top of the vehicle. The FGM-172 is now being fielded in a "B" version that that can be used against buildings and fixed emplacements using a blast-fragmentation multi-purpose warhead. The Predator is soft-launched, which allows the operator to fire the FGM-172 from within an enclosed area without causing harm to himself or his comrades.

GQM-173 MSST

Artist's concept of the GQM-173 in flight.

Specifications

Dimensions: To be Determined
Function: Aerial Target
First Use or Design Date: Under development
Producer: ATK (Alliant Techsystems)
Users: US Navy
Status: Under development

The GQM-173 is a target missile design to mimic the characteristics of the Russian SS-N-27 "Sizzler" anti-ship cruise missile.

This advanced concept is also known as the MSST – or Multi-Stage Supersonic Target, which describes the performance of the missile. As the missile heads towards its target, it cruises as a subsonic vehicle, nearing the target, the first stage drops, leaving the final stage to sprint at supersonic speeds onward.

The GQM-173 is currently in development: details and specifications have not yet been finalized.

RIM-174 ERAM

The RIM-174 leaves the U.S. Navy Arleigh Burke-class guided missile destroyer USS *Dewey* (DDG 105) during initial testing of this latest version of the Standard Missile, the SM-6.

Specifications

Dimensions: To be Determined
Function: Surface-to-Air
Guidance: Radar
First Use or Design Date:
Producer: Raytheon
Users: US Navy
Status: Under development
Family: Standard

The RIM-174 is based on the Standard Missile airframe, but with a seeker head that gives the vehicle the name ERAM, or Extended Range Active Missile. As a derivative of the Standard Missile, it is also known as the SM-6, as the sixth in the series. The active seeker head is from the AIM-120 AMRAAM, and enables the RIM-174 to precisely engage targets at longer ranges. The RIM-174 will replace the RIM-156 in fleet use with AEGIS weapons systems on cruisers and destroyers. Initial RIM-174 production missiles were delivered to the US Navy in 2011.

Missile Name List

AAAM – **AIM-152**
ACM – Raytheon (General Dynamics) **AGM-129**
ADATS – Oerlikon/Lockheed Martin **MIM-146**
AICBM – USAF **BGM-75** (unbuilt)
Agile – Hughes **AIM-95**
ALCM – Boeing **AGM-86**
AMRAAM – Raytheon (Hughes) **AIM-120**
Aquila – Lockheed **MQM-105**
ARCMAT – RS Systems **FQM-117**
ASAT – Vought **ASM-135**
ASRAAM – MBDA (BAe Dynamics/Matra) **AIM-132**
ATACMS – Lockheed Martin (LTV) **MGM-140**
ATACMS II – Lockheed Martin **MGM-164**
ATACMS Block IVA – Lockheed Martin **MGM-168**
Atlas – General Dynamics (Convair) **CGM/HGM-16**
B-Gull – Boeing **GQM-94**
Blue Eye – Martin Marietta **AGM-79**
Bomarc – Boeing **CIM-10**
Bulldog – Texas Instruments **AGM-83**
Bullpup – Martin **AGM-12**
Cardinal – Beech **MQM-61**
Chaparral – Ford **MIM-72**
Chukar – Northrop **MQM/BQM-74**
Condor – Rockwell **AGM-53**
Corporal – JPL/Firestone **MGM-5**
Coyote – Orbital Sciences **GQM-163**
Delta Dagger – General Dynamics/Sperry **PQM-102**
Dragon – McDonnell-Douglas **FGM-77**
EFOGM – Raytheon **MGM-157**
Entac – Aérospatiale (Nord) **MGM-32**
ERAM – Raytheon **RIM-174**
ESSM – Raytheon **RIM-162**
Exdrone – BAI Aerosystems **BQM-147**
Falcon – Hughes **AIM-4**
Falcon – Hughes **AIM-26**
Falcon – Hughes **AIM-47**
Falcon – Hughes **AGM-76**
Falconer – Northrop (Radioplane) **MQM-57**
Firebee – Teledyne Ryan **AQM/BQM/MQM/BGM-34**
Firebolt – Teledyne Ryan **AQM-81**
Firebrand – Teledyne Ryan **BQM-111**
Firefly – Globe **MQM-40**
Firefly – Teledyne Ryan **AQM-91**
Focus – General Electric **AGM-87**
HARM – Raytheon (Texas Instruments) **AGM-88**

Harpoon – Boeing (McDonnell-Douglas) **AGM/RGM/UGM-84**
Have Nap – Rafael/Lockheed Martin **AGM-142**
Hawk – Raytheon **MIM-23**
Hellfire – Boeing/Lockheed Martin (Rockwell/Martin Marietta) **AGM-114**
Hornet – Rockwell (North American) **AGM-64**
Hound Dog – North American **AGM-28**
Hunter – TRW/IAI **BQM-155**
JASSM – Lockheed Martin **AGM-158**
JASSM – Boeing (McDonnell-Douglas) **AGM-159**
Javelin – Raytheon/Lockheed Martin **FGM-148**
JCM – Lockheed Martin **AGM-169** Joint Common Missile
JSOW – Raytheon (Texas Instruments) **AGM-154**
Jupiter – Chrysler **PGM-19**
KEM – Lockheed Martin **MGM-166**
Kingfisher – Lockheed **AQM-60**
Lacrosse – Martin **MGM-18**
Lance – LTV **MGM-52**
LASM – Raytheon **RGM-165**
LOSAT – Lockheed Martin **MGM-166**
Mace – Martin **MGM/CGM-13**
MALD – Northrop Grumman (Teledyne Ryan) **ADM-160**
Matador – Martin **MGM-1**
Mauler – General Dynamics **MIM-46**
Maverick – Raytheon (Hughes) **AGM-65**
Midgetman – Martin Marietta **MGM-134**
Minuteman – Boeing **LGM-30** Minuteman
Minuteman ERCS – Boeing **M-70**
MSST – Alliant Techsystems **MQM-173**
Nike Ajax – Western Electric **MIM-3**
Nike Hercules – Western Electric **MIM-14**
Nike Zeus – Western Electric/McDonnell-Douglas **LIM-49**
Outlaw – Griffon Aerospace **MQM-170**
Outlaw – Griffon Aerospace **MQM-171**
Overseer – Aerojet General **MQM-58**
Patriot – Raytheon **MIM-104**
Pave Cricket – Boeing **CEM-138**
Pave Tiger – Boeing **CQM/CGM-121**
Peacekeeper – Martin Marietta **LGM-118**
Penguin – Kongsberg **AGM-119**
Peregrine – Teledyne Ryan **BQM-145**
Perseus / STAM – **UGM-89** Perseus / STAM
Pershing – Martin Marietta **MGM-31**
Petrel – Fairchild **AQM-41**
Phoenix – Raytheon (Hughes) **AIM-54**
Pointer – AeroVironment **FQM-151**
Polaris – Lockheed **UGM-27**
Poseidon – Lockheed **UGM-73**
Predator – Lockheed-Martin **FGM-172**

Quail – McDonnell **ADM-20**
R-Tern – Teledyne Ryan **GQM-98**
RAM – Raytheon (General Dynamics) **RIM-116**
RCMAT – RS Systems **FQM-117**
Red Head Roadrunner – North American **MQM-42**
Redeye – General Dynamics **FIM-43**
Redstone – Chrysler **PGM-11**
Regulus – Vought **RGM-6**
Regulus II – Vought **RGM-15**
Roland – Euromissile/Hughes/Boeing **MIM-115**
RPVT – USAMICOM **MQM-143**
Sea Lance – Boeing **RUM/UUM-125**
Sea Sparrow – Raytheon **AIM/RIM-7**
Seek Spinner – Boeing **CQM/CGM-121**
Seekbat – General Dynamics **AIM-97**
Sergeant – JPL/Sperry **MGM-29**
Shelduck – Northrop (Radioplane) **MQM-36**
Shillelagh – Ford **MGM-51**
Shrike – Texas Instruments **AGM-45**
Sidearm – Motorola **AGM-122** *S*
Sidewinder – Raytheon (Philco/G.E.) **AIM-9**
Skeeter – Composite Engineering **BQM-167**
Skipper II – Emerson Electric **AGM-123**
Sky Owl – UAV-SR/McDonnell-Douglas **PQM-149**
Skybolt – Douglas **AGM-48**
SLAT – Martin Marietta **AQM-127**
Sparrow – Raytheon **AIM/RIM-7**
Spartan – Western Electric/McDonnell-Douglas **LIM-49**
SRAM – Boeing **AGM-69**
SRAM II – Boeing **AGM-131**
SRAW – Lockheed-Martin **FGM-172**
Standard ARM – General Dynamics **AGM-78**
Standard ER – Raytheon (General Dynamics) **RIM-67**
Standard MR – Raytheon (General Dynamics) **RIM-66**
Standard SM-2ER Block IV – Raytheon **RIM-156**
Standard SM-3 – Raytheon **RIM-161**
Standard ERAM SM-6 – Raytheon **RIM-174**
Streaker – Raytheon (Beech) **MQM-107**
Stinger – Raytheon (General Dynamics) **FIM-92**
Subroc – Goodyear **UUM-44**
Tacit Rainbow – Northrop **AGM/BGM-136**
TALD – IMI (Brunswick) **ADM-141**
Talos – Bendix **RIM-8**
Tartar – General Dynamics (Convair) **RIM-24**
Taurus – (USN) APL **RGM-59** (unbuilt)
Teleplane – USAF FDL **BQM-106**
Terrier – General Dynamics (Convair) **RIM-2**
Thor – Douglas **PGM-17**

Titan – Martin **HGM/LGM-25**
Tomahawk – Raytheon (General Dynamics) **BGM/RGM/UGM-109**
TOW – Raytheon (Hughes) **BGM-71**
Trident I – Lockheed **UGM-96**
Trident II – Lockheed Martin **UGM-133**
TSSAM – Northrop **AGM/MGM-137**
Typhon LR – Bendix **RIM-50**
Typhon MR – Bendix **RIM-55**
Viper – Chrysler **AGM-80**
VL-ASROC – Lockheed Martin (Loral) **RUM-139**
Walleye – Martin Marietta **AGM-62**
Wasp – Hughes **AGM-124**

Nord **MGM-21**
Aérospatiale (Nord) **AGM-22**
Northrop (Radioplane) **MQM-33**
Northrop (Radioplane) **AQM-35**
Beech **AQM-37**
Northrop (Radioplane) **AQM-38**
Beech **MQM-39**
Nord/Bell **PQM-56**
USN **AGM-63** (unbuilt)
USAF **AIM-82** (unbuilt)
USN **RIM-85** (unbuilt)
USN **BQM-90** (unbuilt)
E-Systems **GQM-93**
USA **LIM-99** (unbuilt)
USA **LIM-100** (unbuilt)
USN **RIM-101** (unbuilt)
Teledyne Ryan **AQM-103**
NWC **BQM-108**
LTV **BGM-110**
Rockwell **AGM-112**
NSWC **RIM-113** Anti-Cruise Missile Weapon (unbuilt)
Beech **BQM-126**
USN **AQM-128** Drone (unbuilt)
Boeing (Rockwell) **AGM-130**
DoD **ADM-144** (unbuilt)
PQM-150 *UAV-SR*
USAF **AGM-153** (unbuilt)

Missile Families

ATACMS:
ATACMS – Lockheed Martin (LTV) **MGM-140**
ATACMS II – Lockheed Martin **MGM-164**
ATACMS Block IVA – Lockheed Martin **MGM-168**

Falcon:
Falcon – Hughes **AIM-4**
Falcon – Hughes **AIM-26**
Falcon – Hughes **AIM-47**
Falcon – Hughes **AGM-76**

Sidewinder:
Sidewinder – Raytheon (Philco/G.E.) **AIM-9**
Chaparral – Ford **MIM-72**
Focus – General Electric **AGM-87**
Sidearm – Motorola **AGM-122**

Standard:
Standard ARM – General Dynamics **AGM-78**
Standard ER – Raytheon (General Dynamics) **RIM-67**
Standard MR – Raytheon (General Dynamics) **RIM-66**
Seekbat – Raytheon (General Dynamics) **AIM-97**
Standard SM-2ER Block IV – Raytheon **RIM-156**
Standard SM-3 – Raytheon **RIM-161**
Standard ERAM SM-6 – Raytheon **RIM-174**

Trident:
Trident I – Lockheed **UGM-96**
Trident II – Lockheed Martin **UGM-133**

Bibliography

Collections:

Technical Files
NASM Archives Division
National Air and Space Museum
Smithsonian Institution
Washington, DC

(Note – the NASM Archives Technical Files hold manufacturer trade brochures gathered at various military and defense contractor shows and conventions, or through direct mailings.)

Herbert S. Desind Collection – NASM Acc. No. 1997-0014
NASM Archives Division
National Air and Space Museum
Smithsonian Institution
Washington, DC
(Finding Aid available online at the National Air and Space Museum website:
http://www.nasm.si.edu/research/arch/findaids/pdf/Desind_Finding_Aid.pdf)

Books:

AMI International, Editors
Missiles Systems of the World
Lexington, Massachusetts, Raytheon Company, 1999

Model Designation of Military Aerospace Vehicles
DoD 4120.15-L
Dept. of Defense, Office of the Undersecretary of Defense (Production and Logistics)
Washington, DC, January 1990

Model Designation of Military Aerospace Vehicles
DoD 4120.15-L
Dept. of Defense, Office of the Undersecretary of Defense (Acquisition and Technology)
Washington, DC, March 1996

Model Designation of Military Aerospace Vehicles
DoD 4120.15-L
Dept. of Defense, Office of the Undersecretary of Defense (Acquisition and Technology)
Washington, DC, October 1998

Weapons File – January 1999
Armament Product Group Manager
Eglin AFB, FL, US Air Force, 1999

Aerospace Source Book 2003
Aviation Week and Space Technology – Jan 13, 2003
New York, McGraw-Hill,

Association of Missile and Rocket Industries
US Missiles of All Types Since World War II
Washington, DC, Association of Missile and Rocket Industries 1959

Jane's All The World's Aircraft
Jane's Information Group – various editors
Coulsdon, Surrey, Jane's Information Group (years 1958-2006)

Baker, David
The Rocket: The History and Development of Rocket & Missile Technology
London, New Cavendish Books, 1978

Chant, Christopher
Aircraft Armaments Recognition
Shepperton, Surrey, Ian Allan, Ltd, 1989

Crosby, David F.
A Guide to Airborne Weapons
Charleston, SC, The Nautical & Aviation Publishing Company of America, 2003

Davis, Clive E.
The Book of Missiles
New York, Dodd, Mead & Company, 1959

Elliott, William
Weapons of the US Air Force: A Selective Listing, 1960-2000
HQ Air Force Material Command Office of History, 2000

Gunston, Bill
The Illustrated Encyclopedia of Aircraft Armament
New York, Orion Books, 1988

Jacobs, Horace and Whitney, Eunice Engelke
Missile and Space Projects Guide – 1962
New York, Plenum Press, 1962

Korb, Edward L., Editor
The World's Missile Systems – Eighth Edition, August 1988
Pomona, CA, General Dynamics, 1998

Laur, Timothy M and Llanso, Steven L.
Encyclopedia of Modern US Military Weapons
New York, Berkley Books, 1995

Lennox, Duncan S. and Rees, Arthur, Editors
Jane's Air-Launched Weapons
Coulsdon, Surrey, Jane's Information Group 1990

Merrill, Grayson, Editor
Principles of Guided Missile Design
Princeton, New Jersey, D. Van Nostrand Company, 1959

Munson, Kenneth, Editor
Jane's Unmanned Aerial Vehicles and Targets
Coulsdon, Surrey, Jane's Information Group (years 1999, 2000)

Swanborough, Gordon and Bowers, Peter M.
United States Military Aircraft since 1909
Putnam Aeronautical Books, London, 1989

Swanborough, Gordon and Bowers, Peter M.
United States Navy Aircraft since 1911
Putnam Aeronautical Books, London, 1990

Taylor, John W. R.
Jane's Pocket Book of Remotely Piloted Vehicles
New York, Collier Books 1977

Suggested Further Reading: Selected Histories of Guided Missiles:

Neufeld, Jacob
The Rocket: The History and Development of Rocket & Missile Technology
London, New Cavendish Books, 1978

Ordway, Fredrick I. and Wakeford, Ronald C.
International Missile and Spacecraft Guide
New York, McGraw Hill, 1960

Gunston, William.
The Illustrated Encyclopedia of the World's Rockets and Missiles
New York, Crescent Books, 1979

Werrell, Kenneth
The Evolution of the Cruise Missile
Maxwell Air Force Base, AL, Air University, Air University Press
Washington, DC, US GPO, 1985

Ley, Willy
Rockets, Missiles and Space Travel
New York, Viking, 1961

Gatland, Kenneth William
Development of the Guided Missile
London, Published for Flight by Iliffe, 1954

Stumpf, David K.
Regulus, the Forgotten Weapon
Paducah, KY, Turner Publishing Co. 1997

Smith, Peter Charles
Ship Strike: A History of Air to Sea Weapon Systems
Shrewsbury UK, Airlife, 1998

Chapman, John L.
Atlas: The Story of a Missile
New York, Harper, 1960

Baar, James and Howard, William E.
Polaris!
New York, Harcourt Brace, 1960

Internet Resources as of July, 2009:
http://www.fas.org/man/dod-101/sys/index.html
http://www.globalsecurity.org/military/systems/munitions/missile.htm
http://www.designation-systems.net/dusrm/index.html
http://www.astronautix.com/

Missiles assigned the modern (1962) MDS (Mission Design Series) designations held by the National Air and Space Museum

Martin **MGM-1C** (SSM-A-1 / TM-61C) Matador – UHC
General Dynamics (Convair) **RIM-2A** (SAM-N-7) Terrier – PGF
Western Electric **MIM-3A** (SAM-A-7) Nike Ajax – UHC
Hughes **AIM-4** (GAR-1) Falcon – On Loan
JPL/Firestone **MGM-5** (SSM-A-17) Corporal – UHC
Vought **RGM-6A** (SSM-N-8) Regulus I – UHC
Douglas **AIM -7B** (AAM-N-3) Sparrow II – UHC
Raytheon **AIM -7** Sparrow III –NASM Gallery 203
Bendix **RIM-8J** Talos – UHC
Raytheon (Philco/G.E.) **AIM-9E** Sidewinder – UHC
Raytheon **AIM-9L** Sidewinder (Nose only)-NASM Gallery 203
Chrysler **PGM-11A** (SSM-A-14) Redstone- UHC
Martin **AGM-12A** Bullpup- PGF
General Dynamics (Convair) **CGM-16D** Atlas D – On Loan
General Dynamics (Convair) **HGM-16F** Atlas F – On Loan
Nord **MGM-21A** (SS-10) – PGF
Raytheon **MGM-23A** Hawk – Dulles Storage
Martin **HGM-25A** Titan I – On Loan
Lockheed **UGM-27C** Polaris A-3(Boilerplate) – PGF (Parking Lot)
JPL/Sperry **MGM-29A** Sergeant – Dulles Storage
Boeing **LGM-30G** Minuteman III- NASM Gallery 114
Martin Marietta **MGM-31C** Pershing II – NASM Gallery 100
Aérospatiale (Nord) **MGM-32A** Entac(cutaway) – UHC
Fairchild **AQM-41A** Petrel – On Loan
Goodyear **UUM-44A** Subroc – UHC
Hughes **AIM-54A** Phoenix – NASM Gallery 203
Lockheed **UGM-73A** Poseidon C-3 – UHC
Hughes **AGM-76A** Falcon – UHC
Boeing **AGM-86A** ALCM – UHC
Boeing **AGM-86B** ALCM – UHC
General Dynamics **BGM-109A** Tomahawk – NASM Gallery 114
Boeing/Lockheed Martin **AGM-114C** Hellfire – NASM Gallery 104
Raytheon **AIM-120A** AMRAAM – UHC
Vought **ASM-135A** ASAT – UHC
Raytheon **RIM-161A** Standard SM-3(mockup) – UHC

On Exhibit:
NASM – National Air and Space Museum Downtown Washington, DC
UHC – National Air and Space Museum Udvar-Hazy Center, Chantilly, VA
In Storage:
PGF – National Air and Space Museum, Paul Garber Facility, Suitland, MD
Dulles Storage – NASM Storage Building, Dulles International Airport, VA
(List current as of date of publication)

Index of Designations, Names and Acronyms

AAAM 140
ACM 119
ADATS 135
ADM-20 21
ADM-141 131
ADM-144 133
ADM-160 147
Advanced Air to Air Missile 140
Advanced Cruise Missile 119
Advanced Medium Range Air to Air Missile 110
Agile 87
AGM-12 13, 78
AGM-22 23
AGM-28 29
AGM-45 47, 75, 114
AGM-48 40
AGM-53 55
AGM-62 63
AGM-63 64
AGM-64 64
AGM-65 65, 78
AGM-69 68, 121, 126
AGM-76 73
AGM-78 75, 89
AGM-79 76
AGM-80 76
AGM-83 78
AGM-84 79, 145
AGM-86 80
AGM-87 81
AGM-88 82
AGM-109 98
AGM-112 112
AGM-114 103
AGM-119 109
AGM-122 113
AGM-123 114
AGM-124 115
AGM-129 119
AGM-130 120
AGM-131 121
AGM-136 127
AGM-137 128

AGM-142 132
AGM-153 140
AGM-154 141
AGM-158 145
AGM-159 146
AGM-169 156
AIM-4 4
AIM-7 7, 10, 38, 110
AIM-9 9, 70, 81, 113
AIM-26 27
AIM-47 49, 73
AIM-54 24, 49, 55, 56, 140
AIM-68 67
AIM-82 77
AIM-95 87
AIM-97 89
AIM-120 7, 110
AIM-132 122
AIM-152 140
Air Defense Anti Tank System 135
Air Launched Cruise Missile 80
ALCM 80
AMRAAM 110
Anti Satellite Missile 126
Anti Submarine Rocket 129
AQM-34 35
AQM-35 38
AQM-37 40
AQM-38 41
AQM-41 43
AQM-60 61
AQM-81 77
AQM-91 83
AQM-103 92
AQM-127 118
AQM-128 118
Aquila 94
ARCMAT 107
Army Tactical Missile System 130, 151, 155
ASAT 126
ASM-135 126
ASRAAM 132

ASROC 129
ATACMS 130
ATACMS II 151
ATACMS Block IVA 155
Atlas 17
Augmented Radio Controlled Miniature Aerial Target 107

B-Gull 86
BGM-34 35
BGM-71 69
BGM-109 98
BGM-110 100
BGM-136 127
Blue Eye 76
Bomarc 11, 61
BQM-34 35, 117
BQM-74 72
BQM-90 83
BQM-106 95
BQM-108 97
BQM-111 101
BQM-126 117
BQM-145 134
BQM-147 136
BQM-155 142
BQM-167 154
Brazo 64
BroadSword 158
Bulldog 78
Bullpup 13, 78

Cardinal 42, 62
CEM-138 128
CGM-13 14
CGM-16 17
CGM-121 112
Chaparral 70
Chukar 72
CIM-10 11, 61
Compass Cope 89
Compass Cope B 86
Condor 55
Corporal 5, 30
Coyote 150

Creeper Prop 34
CT.41 58
CQM-10 11
CQM-121 112

Delta Dagger 91
Dogfight 77
Dragon drone 136
Dragon missile 74

EFOGM 144
Engin Téléguidé Anti-Char 33
Enhanced Fiber Optic Guided Missile 144
Entac 33
ERAM 161
ESSM 149
Evolved Sea Sparrow Missile 149
Exdrone
Extended Range Active Missile 161

Falcon 4, 27, 49, 73
Falconer 59
FGM-77 74
FGM-148 137
FGM-172 159
FIM-43 45, 84
FIM-92 84
Firebee 35, 92, 117
Firebolt 77
Firebrand 101
Firefly 83
Focus 81
FQM-117 107
FQM-151 139

GQM-93 85
GQM-94 86
GQM-98 86, 89
GQM-163 150
GQM-173 160
Gryphon 98

HARM 82
Harpoon 79
Have Nap 132

Hawk 24, 44
Hellfire 103
HGM-16 17
HGM-25 26
High-speed Anti Radiation Missile 82
Hornet 64
Hound Dog 29
Hunter 142

Improved Tactical Air Launched Decoy 131
ITALD 131

JASSM 145, 146
Javelin 137
Jayhawk 40
JCM 156
Joint Air to Surface Standoff Missile 145, 146
Joint Common Missile 156
Joint Stand Off Weapon 141
JSOW 141
Jupiter 20

KEM 153
Kinetic Energy Missile 153
Kingfisher 61

Lacrosse 19, 54
Lance 30, 54
Land Attack Standard Missile 152
LASM 152
LGM-25 26
LGM-30 31
LGM-118 31, 108
LIM-49 51
LIM-99 90
LIM-100 90
Line-Of-Sight Anti-Tank Weapon 153
LOSAT 153

M-70 68
M-75 72
Mace 14
MALD 147

Matador 1
Mauler 48, 70
Maverick 65, 78
MGM-1 1
MGM-5 5, 30
MGM-13 14
MGM-18 19
MGM-21 22
MGM-29 5, 30
MGM-31 32
MGM-32 33
MGM-51 53
MGM-52 30, 54
MGM-134 125
MGM-137 128
MGM-140 130, 151, 155
MGM-157 144
MGM-164 151
MGM-166 153
MGM-168 155
Midgetman 125
MIM-3 3, 15
MIM-14 3, 15, 38, 61
MIM-23 24, 44
MIM-46 48, 70
MIM-72 70
MIM-104 93
MIM-115 105
MIM-146 135
Miniature Air Launched Decoy 147
Minuteman 31
MQM-8 8
MQM-33 34, 39, 59
MQM-34 35
MQM-36 39
MQM-39 42
MQM-40 42
MQM-42 44
MQM-57 59, 60
MQM-58 60
MQM-61 62
MQM-74 72, 97
MQM-105 94
MQM-107 96
MQM-143 133
MQM-170 157, 158
MQM-171 158

Index

MR UAV 134
MSST 160
Multi Stage Supersonic Target 160
MX 108

Nike Ajax 3, 15
Nike Hercules 3, 15, 38, 51, 61
Nike Zeus 51

Outlaw 157
Overseer 60

Patriot 93
Pave Arm 64
Pave Cricket 128
Pave Tiger 112
Peacekeeper 31, 108
Penguin 109
Peregrine 134
Pershing 32
Petrel 43
Phoenix 24, 49, 55, 56, 140
PGM-11 12, 20, 32
PGM-17 18
PGM-19 20
Pointer 139
Polaris 28
Popeye 132
Poseidon 71
PQM-56 58
PQM-102 91
PQM-149 138
PQM-150 138
Predator 159

Quail 21

R-Tern 89
Radio Controlled Miniature Aerial Target 107
RAM 106
Raptor 132
RCMAT 107
Red Head Roadrunner 44
Redeye 45, 84
Redstone 12, 20, 32

Regulus 6, 16
Regulus II 16
Remotely Piloted Vehicle Target 133
RGM-6 6
RGM-15 16
RGM-59 60
RGM-84 79
RGM-109 98
RGM-165 152
RIM-2 2, 67
RIM-7 7
RIM-8 8
RIM-24 25, 66
RIM-50 52
RIM-55 57
RIM-66 25, 66
RIM-67 67
RIM-85 79
RIM-101 90
RIM-113 102
RIM-116 116
RIM-156 143
RIM-161 148
RIM-162 149
RIM-174 161
Roland 105
Rolling Airframe Missile 106
RPVT 133
RQ-5 142
RUM-125 116
RUM-139 129

Sea Lance 116
Sea Sparrow 7
Seek Spinner 112
Seekbat 89
Sergeant 5
Shelduck 39
Shillelagh 53
Shrike 47, 75
Short Range Attack Missile 68, 121, 126
Short Range Attack Missile II 121
Short Range Attack Weapon 159
SICBM 125

Sidearm 113
Sidewinder 9, 70, 81, 87, 106, 113
Skeeter 154
Skipper II 114
Sky Owl 138
Skybolt 50
SLAT 118
SNIPE 63
Sparrow 7, 10, 38, 110
Spartan 51
SRAM 68, 121, 126
SRAM II 121
SRAW 159
SS.10 22
SS.11 23
SSST 150
Standard ARM 75, 89
Standard ER 66, 67
Standard MR 25, 66
Standard SM-2ER Block IV 143
Standard SM-3 148
Standard ERAM SM-6 161
Streaker 96
Stinger 84, 106
Supersonic Low Altitude Target 118
Supersonic Sea Skimming Target 150
Subroc 46

Tacit Rainbow 127
Tactical Air Launched Decoy 131
TALD 131
Talos 8, 52
Tartar 25, 52, 66
Teleplane 95
Terrier 2, 52, 60, 67
Thor 18
Titan 26
Tomahawk 98
TOW 69
Tri Service Standoff Attack Missile 128
Trident I 71, 88, 123
Trident II 88, 123

TSSAM 128
Tube launched Optically tracked
    Wire guided missile 69
Typhon LR 52
Typhon MR 57

UAV-SR 138
UGM-27 28
UGM-73 71

UGM-84 79
UGM-89 82
UGM-96 88, 123
UGM-109 98
UGM-133 88, 123
UUM-44 46
UUM-125 116

Vandal 8

Vertical Launched Anti
    Submarine Rocket 129
Viper 76
VL-Asroc 129

Walleye 63
Wasp 115

Photo Credits

Page 1 TL, US Air Force via National Air and Space Museum (NASM 9A03291); Smithsonian Institution. (Hereafter by NASM number); Page 1 TR, US Air Force (NASM 9A03292); Page 2 TL, US Navy (NASM 9A03293); Page 2 TR, Convair (NASM 9A03294); Page 3 TL, US Army (NASM 9A03295); Page 3 TR, (NASM 9A03296); Page 4 TL, US Air Force (NASM 9A03297); Page 4 TR, (NASM 9A03298); Page 5 TL US Army (NASM 9A03299); Page 5 TR, US Steel photo (NASM 9A03356); Page 6 TL US Navy (NASM 9A03357); Page 6 TR, US Navy (NASM 9A03358); Page 7 TL (NASM 9A03359); Page 7 TR, US Navy (NASM 9A03360); Page 8 TL, US Navy (NASM 9A03361); Page 8 TR, US Navy (NASM 9A03362); Page 9 TL, Raytheon; Page 9 TR (NASM 9A03363); Page 9 BL, US Navy; Page 9 BR, US Air Force (NASM 9A03364); Page 11 TL (NASM 9A03365); Page 11 TR, US Air Force (NASM 9A03366); Page 12 TL, US Army (NASM 9A03367); Page 12 TR, US Army (NASM 9A03368); Page 13 TL (NASM 9A03369); Page 13 TR, US Navy (NASM 9A03370); Page 14 TL (NASM 9A03371); Page 14 TR, Lockheed-Martin (NASM 9A03372); Page 15 TL, DoD (NASM 9A03373); Page 15 TR, US Army (NASM 9A03374); Page 16 TL, US Navy (NASM 9A03375); Page 16 TR, US Air Force (NASM 9A03376); Page 17 TL, US Air Force (NASM 9A03377); Page 17 TR, US Air Force (NASM 9A03378); Page 18 TL (NASM 9A03379); Page 18 TR, US Air Force (NASM 9A03380); Page 19 TL, US Army (NASM 9A03381); Page 19 TR, US Army (NASM 9A03382); Page 20 TR, US Air Force (NASM 9A03383); Page 20 BR, US Air Force (NASM 9A01841); Page 21 TL, Boeing Photo (McDonnell Douglas D4C-8949) (NASM 9A03385); Page 21 TR, US Air Force (NASM 9A03384); Page 21 BR, Boeing Photo (McDonnell Douglas D4C-3930) (NASM 9A03386); Page 22 TL, US Army (NASM 9A03387); Page 22 TR Aerospatiale/Nord Photo (NASM 9A03388); Page 23 TL, US Army (NASM 9A03389); Page 23 TR, US Army; Page 23 BR Bell Helicopter Textron via (NASM 9A03390); Page 24 TL BAE Systems/FMC Corp, Ordnance Div photo 7582 (NASM 9A03391); Page 24 TR Raytheon Photo (NASM 9A03392); Page 25 TL, US Navy (NASM 9A03393); Page 25 TR (NASM 9A01843); Page 25 BR, US Navy (NASM 9A01844); Page 26 TL , US Air Force (NASM 9A03394); Page 26 TM (NASM 9A03395); Page 26 TR, US Air Force; Page 27 TL Hughes Aircraft photo (NASM 9A01887); Page 27 TR Hughes Aircraft photo (NASM 9A01888); Page 28 TL (NASM 9A03396); Page 28 TR US Navy (NASM 9A03397); Page 28 BR US Navy (NASM 9A03398); Page 29 TL, US Air Force (NASM 9A03399); Page 29 TR, US Air Force (NASM 9A01700); Page 30 TL, US Army (NASM 9A01701); Page 30 TR, US Army (NASM 9A01702); Page 31 TL, US Air Force (NASM 9A01703); Page 31 TR (SI - NASM 98-15302); Page 32 TL, Lockheed-Martin (NASM 9A01705); Page 32 TR (NASM 9A01706); Page 33 TL (NASM 9A01707); Page 33 TR (NASM 9A01708); Page 34 TL, Northrop-Grumman (NASM 9A01849); Page 34 TR, Northrop-Grumman (NASM 9A01850); Page 34 BR, Northrop-Grumman (NASM 9A01851); Page 35 TL, Northrop-Grumman (NASM 9A01709); Page 35 TR, US Navy (NASM 9A01710); Page 35 BR (NASM 9A04355); Page 36 TL, US Army (NASM 9A04357); Page 36 TR (NASM 9A04358); Page 36 BL, US Air Force (NASM 9A04359); Page 36 BR (NASM 9A04360); Page 37 TL, Northrop-Grumman (NASM 9A04361); Page 37 TR, Northrop-Grumman (NASM 9A04362); Page 37 BL, US Air Force (NASM 9A04363); Page 37 BR (NASM 9A04356); Page 38 TL, US Air Force (NASM 9A01852); Page 38 TR, US Air Force (NASM 9A01853); Page 38 BL, Northrop-Grumman (NASM 9A01854); Page 38 BR, US Air Force (NASM 9A01855); Page 39 TL, Northrop-Grumman (NASM 9A01856); Page 39 TR (NASM 9A01857); Page 40 TL, DoD (NASM 9A01711); Page 40 TR, Beech; Page 41 TL, Northrop-Grumman (NASM 9A01889); Page 41 TR, Northrop-Grumman (NASM 9A01890); Page 41 BR, Northrop-Grumman (NASM 9A01891); Page 42 TL, Beech (NASM 9A01892); Page 42 TR (NASM 9A01893); Page 43 TL US Navy (NASM 9A01894); Page 43 TR (NASM 9A01895); Page 43 BR (SI-NASM 87-13080); Page 44 TL, US Army (NASM 9A04333); Page 44 TR (NASM 9A04334); Page 44 BL (NASM 9A04335); Page 44 BR (NASM 9A04336); Page 45 TL, US Army 9A01846); Page 45 TR, US Army (NASM 9A01847); Page 45 BR, US Army (NASM 9A01848); Page 46 TL Goodyear Aerospace Corporation N63091329 (NASM 7B30356); Page 46 TR Goodyear Aerospace Corporation N66092917 (NASM 9A30383); Page 46 BR Goodyear Aerospace Corporation N62071104 (NASM 9A30384); Page 47 TL, US Air Force (NASM 9A01712); Page 47 TR, Texas Instruments (NASM 9A01713); Page 48 TL, BAe (NASM 9A01714); Page 48 TR, US Army (NASM 9A01715); Page 48 BR, US Army (NASM 9A01716); Page 49 TL, Hughes (NASM 7B23985); Page 49 TR, NASA (NASM 7B14877); Page 50 TL (NASM 9A01717); Page 50 TR, US Air Force (NASM 9A01718); Page 51 TL, US Army (NASM 9A01719); Page 51 TR, US Army (NASM 9A01720); Page 51 BR, US Army (NASM 9A01721); Page 52 TL, US Navy (NASM 9A01722); Page 52 TR, US Navy (NASM 9A01723); Page 53 TL, US Army (NASM 9A01725); Page 53 TR, Philco (NASM 9A01726); Page 53 BR, US Army (NASM 9A01724); Page 54 TL, US Army (NASM 9A01727); Page 54 TR (NASM 9A01728); Page 54 BR (NASM 9A01730); Page 55 TL, US Navy (NASM 9A01732); Page 55 TR, US Navy (NASM 9A01731); Page 55 BR, US Navy (NASM 9A01733); Page 56 TL, Hughes (NASM 9A01735); Page 56 TR, US Navy (NASM 9A01734); Page 58 TL (NASM 9A01899); Page 58 TR (NASM 9A04332); Page 58 BR Aerospatiale/Nord Photo 41-12/11 (NASM 9A01898); Page 59 TL (NASM 9A01858); Page 59 TR (NASM 9A01859); Page 60 TL (NASM 9A01860); Page 60 TR (NASM 9A01861); Page 61 TL, Lockheed-Martin (NASM 9A01863); Page 61 TR (NASM 9A01862); Page 62 TL (NASM 9A01896); Page 62 TR, Beech (NASM 9A01897); Page 63 TL, US Navy (NASM 9A01736); Page 63 TR, US Navy (NASM 9A01737); Page 64 (NASM 9A01840); Page 65 TL, Hughes (NASM 9A01738); Page 65 TR, Hughes (NASM 9A01739); Page 66 TL, US Navy (NASM 9A04348); Page 66 TR, US Navy (NASM 9A04349); Page 67 TL, US Navy (NASM 9A04350); Page 67 TM, US Navy (NASM 9A04351) , Page 67 TR, DoD (NASM 9A04352); Page 68 TL, US Air Force (NASM 9A01741); Page 68 TR (NASM 9A01742); Page 69 TL, Hughes (NASM 9A01743); Page 69 TR, Hughes via DoD (NASM 9A01744); Page 70 TL (NASM 9A01745); Page 70 TR, Philco (NASM 9A01746); Page 70 BR (NASM 9A01829); Page 71 TL, US Air Force (NASM 9A01830); Page 71 TR, US Air Force (NASM 9A01831); Page 72 TL (NASM 9A01837); Page 72 TR (NASM 9A01839); Page 73 TL US Navy Photo, Page 73 TR US Navy Photo; Page 74 TL (NASM 9A01750); Page 74 TM Boeing Photo (McDonnell Douglas

12-351-5) (NASM 9A01748); Page 74 TR, US Navy (NASM 9A01749); Page 75 TL, US Air Force (NASM 9A01751); Page 75 TR Boeing Photo (McDonnell Douglas C22-287-2 dated 2/82) (NASM 9A01845); Page 76, via National Archives (NASM 9A01752); Page 77, DoD (NASM 9A01753); Page 78 TL, US Navy (NASM 9A01756); Page 78 TR, US Navy (NASM 9A01755); Page 78 BR, US Navy (NASM 9A01754); Page 79 TL, Boeing (McDonnell-Douglas) (NASM 9A01825); Page 79 TR, US Navy (NASM 9A01826); Page 80 T, US Air Force (NASM 9A01757); Page 80 BR, US Air Force (NASM 9A01758); Page 81 TL, US Air Force (NASM 9A02222); Page 81 BR, US Air Force via DoD (NASM 9A01759); Page 82 TL, US Air Force (NASM 9A01761); Page 82 TR, Texas Instruments (NASM 9A01827); Page 82 BR (NASM 9A01828); Page 83 ML (NASM 9A01874); Page 83 MR (NASM 9A01880); Page 84 TL, General Dynamics Pomona (NASM 9A01763); Page 84 TR (NASM 9A01762); Page 85 TR courtesy of Jun Shidara; Page 86 TL (NASM 9A01875); Page 86 TR (NASM 9A01876); Page 86 BR (NASM 9A01877); Page 87 TL, US Navy (NASM 9A01765); Page 87 TR, US Navy (NASM 9A01764); Page 87 BR, US Navy via Carl Cline; Page 88 TL, US Navy (NASM 9A01866); Page 88 TR, US Air Force (NASM 9A01864); Page 88 BR, US Air Force (NASM 9A01865); Page 89 TR, US Air Force (NASM 9A01766); Page 89 BL, Teledyne Ryan (NASM 9A01878); Page 89 BR (NASM 9A01879); Page 91 TL courtesy of Joe Handelman (NASM 9A01767); Page 91 TR (NASM 9A01768); Page 92 TL courtesy Northrop-Grumman; Page 92 TR courtesy Northrop-Grumman; Page 93 TL (NASM 9A01747); Page 93 TR (NASM 9A01769); Page 93 BR (NASM 9A04330); Page 94 TL (NASM 9A01770); Page 94 TR (NASM 9A01771); Page 94 BR (NASM 9A01772); Page 95 ALL courtesy USAF Flight Dynamics Laboratory, Page 96 TL, US Army (NASM 9A01773); Page 96 TR (NASM 9A01774); Page 97 T courtesy USN Carderock Division, Naval Surface Warfare Center; Page 98 TL (NASM 9A01775); Page 98 TR (NASM 9A01776); Page 98 BL, DoD (NASM 9A01777); Page 98 BR, DoD (NASM 9A01778); Page 99 TL, DoD (NASM 9A01779); Page 99 TR, DoD (NASM 9A01780); Page 99 BR, DoD (NASM 9A01781); Page 100 T, US Navy (NASM 9A01882); Page 101 TL US Navy (NASM 9A01881); Page 101 TR (San Diego Aerospace Museum); Page 101 BR (San Diego Aerospace Museum); Page 102 TL US Air Force; Page 102 TR US Air Force; Page 103 TL (NASM 9A01782); Page 103 TR, Boeing (McDonnell-Douglas) (NASM 9A01783); Page 103 B, Boeing (McDonnell-Douglas) (NASM 9A01785); Page 105 TL (NASM 9A01786); Page 105 TR (NASM 9A01787); Page 105 BR (NASM 9A01788); Page 106 TL, General Dynamics Pomona (NASM 9A01789); Page 106 TR, General Dynamics Pomona (NASM 9A01790); Page 106 BR (NASM 9A01791); Page 107 T, RS Systems (NASM 9A04347); Page 108 TL (NASM 9A01792); Page 108 TR (NASM 9A01793); Page 108 BR, US Air Force; Page 109 TL (NASM 9A01794); Page 109 BL (NASM 9A01796); Page 109 BR (NASM 9A01797); Page 110 TL, Hughes (NASM 9A04331); Page 110 TR, Hughes (NASM 9A01832); Page 110 BL, US Air Force via DoD (NASM 9A01834); Page 110 BR, US Air Force via DoD (NASM 9A01833); Page 111 TL, Hughes (NASM 9A01835); Page 111 TR, Hughes (NASM 9A01836); Page 111 BR, DoD; Page 112 TL Boeing photo (BMAC 287-640) (NASM 9A01798); Page 112 TR (NASM 9A01799); Page 112 BR (NASM 9A01800); Page 113 TL, US Navy (NASM 9A04323); Page 113 TR, Boeing (McDonnell-Douglas Helicopter) (NASM 9A04322); Page 114 TL; Page 114 TR, US Air Force; Page 115 T, Hughes (NASM 7B34616); Page 116 TL Boeing photo (BAC B-1024 83SK04965-5) (NASM 9A04337); Page 116 TR Boeing photo (BAC B-0943 83SK04965-3) (NASM 9A04338); Page 116 BR Boeing photo (BAC B-1236 R2588) (NASM 9A04339); Page 117 T (NASM 9A01801); Page 118 TL US Navy Photo; Page 118 TR US Navy Photo; Page 118 BR US Navy Photo; Page 119 TL (NASM 9A01802); Page 119 TR, US Air Force (NASM 9A01803); Page 120 TL US Air Force; Page 120 TR US Air Force; Page 120 BL, DoD; Page 121 TL, Boeing (NASM 9A01804); Page 121 TR, Boeing (NASM 9A01805); Page 122 TL, BAe (NASM 9A01807); Page 122 TR, BAe (NASM 9A01806); Page 123 TL, US Navy (NASM 9A01869); Page 123 BL, Lockheed-Martin (NASM 9A01867); Page 123 BR, DoD (NASM 9A01868); Page 124 BL, US Air Force (NASM 9A01870); Page 125 TL, Lockheed-Martin (NASM 9A01872); Page 125 BR, US Air Force (NASM 9A01873); Page 126 TL, US Air Force (NASM 7B10820); Page 126 TR (NASM 9A01808); Page 126 BR (NASM 9A01809); Page 127 TL, Northrop-Grumman (NASM 9A01810); Page 127 TR (NASM 9A01811); Page 128 TL, Northrop-Grumman; Page 129 TL, US Navy (NASM 9A04354); Page 129 TR, US Navy (NASM 9A04353); Page 129 BR, US Navy; Page 130 TL (NASM 9A01812); Page 130 TR (NASM 9A01813); Page 131 TL (NASM 9A01883); Page 131 TR (NASM 9A01884); Page 131 BL (NASM 9A01885); Page 131 BR, Intellitec Brunswick Defense (NASM 9A01886); Page 132 TL, Lockheed-Martin (NASM 9A01814); Page 132 TR, DoD; Page 132 BR, US Air Force; Page 134 TL, Northrop-Grumman; Page 135 TL (NASM 9A04340); Page 135 TR (NASM 9A04342); Page 135 BR (NASM 9A04341); Page 136 TL photo by Tien-Seng Chin, courtesy of L-3 BAI Aerosystems; Page 137 TL, Texas Instruments (NASM 9A04344); Page 137 TR, Hughes (NASM 9A04343); Page 137 BL, Raytheon (NASM 9A04366); Page 139 TL (NASM 9A01815); Page 139 TR (NASM 9A01816); Page 139 BR (NASM 9A01817); Page 140 T, Hughes (NASM 9A04346); Page 140 BR, Hughes (NASM 9A04345); Page 141 TL, Raytheon (NASM 9A04368); Page 141 TR, Raytheon (NASM 9A04369); Page 141 BL, US Air Force; Page 142 TL (NASM 9A01818); Page 142 TR (NASM 9A01819); Page 143 TL, Raytheon (NASM 9A04370); Page 144 TL, Raytheon (NASM 9A04371); Page 144 TR, Steve Zaloga; Page 144 BR, US Army; Page 145 TL, US Air Force; Page 145 TR, US Air Force; Page 145 BL, US Air Force; Page 146 TL, Boeing Photo (McDonnell Douglas); Page 146 BL, Boeing Photo (McDonnell Douglas); Page 147 TL, Northrop-Grumman; Page 147 TR, Raytheon ; Page 147 B, Steve Zaloga; Page 148 TL, US Navy; Page 148 BR, Raytheon (NASM 9A04372); Page 149 TL, Raytheon (NASM 9A04373); Page 149 TR, Raytheon (NASM 9A04374); Page 150 TL, US Navy via OSC (NASM 9A01820); Page 150 TR, US Navy via OSC (NASM 9A01821); Page 151 TL, Lockheed-Martin ; Page 151 TR, Lockheed-Martin; Page 151 BR, Lockheed-Martin; Page 152 TL, Raytheon (NASM 9A04375); Page 153 TL Northrop Grumman (LTV photo) (NASM 9A01842); Page 153 TR, Lockheed-Martin (NASM 9A01822); Page 154 TL, US Air Force; Page 154 TR, US Air Force; Page 154 B, Steve Zaloga; Page 155 T Lockheed-Martin; Page 156 TL, Lockheed-Martin (NASM 9A01823); Page 156 TR, Lockheed-Martin (NASM 9A01824); Page 157 TL Griffon Aerospace; Page 157 TR Griffon Aerospace; Page 157 BR Griffon Aerospace; Page 158 TL Griffon Aerospace; Page 158 TR Griffon Aerospace; Page 159 TL Lockheed-Martin; Page 159 TR Lockheed-Martin; Page 160 TL ATK; Page 161 TR Raytheon.